Biological Effects of DDT in Lower Organisms

Papers by
Maria Luisa Dinamarca, Frederick W. Plapp, Jr.,
B.E. Langlois et al.

IN COOPERATION WITH THE
SMITHSONIAN SCIENCE INFORMATION EXCHANGE

Summaries of current research projects are included in the
final section of this volume. Previously unpublished, these
summaries were obtained from a search conducted by the
Smithsonian Science Information Exchange, a national collection of information on ongoing and recently terminated
research.

MSS Information Corporation
655 Madison Avenue, New York, N.Y. 10021

71001

Library of Congress Cataloging in Publication Data
Main entry under title:

Biological effects of DDT in lower organisms.

    1. DDT (Insecticide)--Toxicology--Addresses, essays,
lectures. 2. DDT (Insecticide)--Physiological effect--
Addresses, essays, lectures. I. Dinamarca, Maria
Luisa.
RA1242.D35B56   574.2'4   73-12476
ISBN 0-8422-7120-1

TABLE OF CONTENTS

CREDITS AND ACKNOWLEDGEMENTS

Atallah, Y.H., "Rate of Increased Tolerance to Insecticides in the Egyptian Cotton Leafworm, *Spodoptera littoralis* (Boisd.)," *Zeitschrift für Angewandte Entomologie*, 1972, 71:6-12.

Atallah, Yousef H., "Status of Carbaryl. and DDT Resistance in Laboratory-Reared Egyptian Cotton Leafworm," *Journal of Economic Entomology*, 1971, 64:1018-1021.

Axtell, R.C.; and T.D. Edwards, "Susceptibility of Adult *Hippelates pusio* to Insecticidal Fogs," *Journal of Economic Entomology*, 1970, 63:1184-1185.

Bahr, Thomas G.; and Robert C. Ball, "Action of DDT on Evoked and Spontaneous Activity from the Rainbow Trout Lateral Line Nerve," *Comparative Biochemistry and Physiology*, 1971, 38A:279-284.

Bitman, Joel; Helene C. Cecil; Susan J. Harris; and George F. Fries, "Comparison of DDT Effect on Pentobarbital Metabolism in Rats and Quail," *Agricultural and Food Chemistry*, 1971, 19:333-338.

Bitman, Joel; Helene C. Cecil; and George F. Fries, "DDT-Induced Inhibition of Avian Shell Gland Carbonic Anhydrase: A Mechanism for Thin Eggshells," *Science*, 1970, 168: 594-596.

Chopra, N.M.; and Neil B. Osborne, "Systematic Studies on the Breakdown of p,p'-DDT in Tobacco Smokes. II. Isolation and Identification of Degradation Products from the Pyrolysis of p,p'-DDT in a Nitrogen Atmosphere," *Analytical Chemistry*, 1971, 43:849-853.

Cox, James L., "Accumulation of DDT Residues in *Triphoturus mexicanus* from the Gulf of California," *Nature*, 1970, 227:192-193.

Dinamarca, Maria Luisa; Leo Levenbook; and Elena Valdés, "DDT-dehydrochlorinase. II. Subunits, Sulfhydryl Groups, and Chemical Composition," *Archives of Biochemistry and Biophysics*, 1971, 147:374-383.

Donaldson, W.E.; M.D. Jackson; and T.J. Sheets, "Influence of Iodinated Casein on DDT Residues in Chicks," *Poultry Science*, 1971, 50:1316-1320.

Edland, Torgeir, "Laboratory and Field Evaluation of the Persistence of Some Insecticides on Noctuid Larvae on Apple in Norway," *Journal of Economic Entomology*, 1972, 65: 208-211.

French, Allen L.; and Roger A. Hoopingarner, "Dechlorination of DDT by Membranes Isolated from *Escherichia coli*," *Journal of Economic Entomology*, 1970, 63:756-759.

Hill, Elwood F.; William E. Dale; and James W. Miles, "DDT Intoxication in Birds: Subchronic Effects and Brain Residues," *Toxicology and Applied Pharmacology*, 1971, 20: 502-514.

Janicki, Ralph H.; and William B. Kinter, "DDT: Disrupted Osmoregulatory Events in the Intestine of the Eel *Anguilla rostrata* Adapted to Seawater," *Science*, 1971, 173:1146-1148.

Keiser, Irving; Esther L. Schneider; and Isao Tomikawa, "Species Specificity among Oriental Fruit Flies, Melon Flies, and Mediterranean Fruit Flies in Susceptibility to Insecticides at Several Loci," *Journal of Economic Entomology*, 1971, 64:606-610.

Kutches, A.J.; and D.C. Church, "DDT-[14]C Metabolism by Rumen Bacteria and Protozoa *in Vitro*," *Journal of Dairy Science*, 1971, 54:540-543.

Langlois, B.E.; J.A. Collins; and K.G. Sides, "Some Factors Affecting Degradation of Organochlorine Pesticides by Bacteria," *Journal of Dairy Science*, 1970, 53:1671-1675.

Malone, Thomas C., "*In Vitro* Conversion of DDT to DDD by the Intestinal Microflora of the Northern Anchovy, *Engraulis mordax*," *Nature*, 1970, 227:848-849.

McLennan, D.J.; and R.J. Wong, "The Mechanism of the Chloride Ion-Promoted Dehydrochlorination of DDT and Ring-Substituted Analogues in Acetone," *Tetrahedron Letters*, 1970, No. 11:881-884.

Onsager, Jerome A.; H.W. Rusk; and L.I. Butler, "Residues of Aldrin, Dieldrin, Chlordane, and DDT in Soil and Sugarbeets," *Journal of Economic Entomology*, 1970, 63:1143-1146.

Peaslee, Margaret H., "Influence of DDT upon Pituitary Melanocyte-Stimulating Hormone (MSH) Activity in the Anuran Tadpole," *General and Comparative Endocrinology*, 1970, 14:594-595.

Plapp, Frederick, Jr.; and John Casida, "Induction by DDT and Dieldrin of Insecticide Metabolism by House Fly Enzymes," *Journal of Economic Entomology*, 1970, 63:1091-1092.

Plapp, Frederick W., Jr., "Changes in Glucose Metabolism Associated with Resistance to DDT and Dieldrin in the House Fly," *Journal of Economic Entomology*, 1970, 63:1768-1772.

PREFACE

DDT is now known to interact with living organisms at all levels of biological organization. The present collection of papers, published from 1970 to 1972, documents the effects of DDT in lower vertebrates and invertebrates. Biodegradation of DDT in microorganisms and in insect tissues is considered in biochemical detail in the first section, while DDT effects in birds are discussed in later sections. New procedures for chemical analysis of DDT and its derivatives are described, and the phenomenon of DDT resistance in insects is evaluated. This collection is a companion volume to *Effects of DDT on Man and Other Mammals, Vols. I & II* also published by MSS, and together they should provide comprehensive analysis of the effects of DDT on all living organisms.

Biodegradation of DDT by Insects and Microorganisms

DDT-dehydrochlorinase II.

Subunits, Sulfhydryl Groups, and Chemical Composition

MARIA LUISA DINAMARCA, LEO LEVENBOOK,[1] AND ELENA VALDÉS

*Department of Chemistry and Biochemistry, University of Chile School of Medicine, Santiago,
Chile, and National Institute of Arthritis and Metabolic Diseases, National Institutes
of Health, Bethesda, Maryland 20014*

We have previously described the purification and some of the properties of DDT-dehydrochlorinase[2] (EC 4.5.1.1, hereafter referred to as DDT-ase) isolated from a DDT-resistant strain of houseflies (1). Our data suggested that the enzyme consisted of four subunits of mol wt $\sim$30,000, which in the presence of substrate DDT, became aggregated into a tetrameric structure of mol wt $\sim$120,000 having maximal enzymatic activity. The trimer was only 10 % as active, while dimers and monomers were apparently inactive. The four monomers were inseparable by either gel filtra-

tion or by sucrose-density centrifugation, and hence by these procedures it was impossible to ascertain whether they are dissimilar or identical. We now present chemical data bearing on this question as well as on the importance of protein thiol groups in relation to enzymatic activity.

MATERIALS AND METHODS

Materials. Houseflies (*Musca domestica* L.), strain P_2/sel, were reared as previously described (1); about 5000 nonsexed adult insects, 5 days postemergence, were employed for a single preparation of the enzyme.

Twice recrystallized trypsin, ultrapure grade urea, and DNP–amino acid standards were purchased from Mann Research Laboratories, N.Y. GSH and PCMB, Na-salt, were obtained from Sigma Chemical Co., St. Louis. DDT (Fisher Scientific Co.) was purified and assayed as described (1). Dithiothreitol was kindly donated by Dr. H. Gelboin. Mercaptoethanol was purchased

[1] To whom further communication should be addressed at the National Institutes of Health.

[2] Abbreviations: DDT, 1,1,1-trichloro-2,2-bis-(*p*-chlorophenyl) ethane; PCMB, *p*-chloromercuribenzoate; DDE, 1,1-dichloro-2,2-bis(*p*-chlorophenyl) ethylene; GSH, reduced glutathione; SDS, sodium dodecyl sulfate.

from Light and Co., Colnbrook, England, FDNB from Baker Chemical Co., N.J., and the gel-electrophoresis apparatus and reagents from Canalco, Rockville, MD.

Enzyme purification. DDT-ase was prepared by modification and simplification of the previously published procedure (1). These improvements resulted in a 2- to 3-fold increase in yield of enzyme of about the same specific activity. The modifications were as follows. (A) All procedures were performed at 2–4°, and all buffers except where otherwise noted, contained either 0.01 M GSH or mercaptoethanol. Addition of the thiol reagent ensured isolation of the enzyme in its monomeric form. (B) After dialysis of fraction II as described (1), the extract was lyophilized, dissolved in a minimal volume of 0.01 M phosphate buffer, pH 7.4, and passed over a 1 × 30-cm column of Sephadex G-50 equilibrated with the above which was also used to elute the protein in the void volume. This step served to eliminate several low molecular weight pigments. (C) The eluate was concentrated by lyophilization and applied to a 2.5 × 40-cm column of Sephadex G-100. One-milliliter fractions were collected at a flow rate of 0.3 ml/min, and a typical result is depicted in Fig. 1A. The protein fractions demonstrating enzyme activity (peak No. 3) were combined, and the

buffer GSH (or mercaptoethanol) removed by dialysis against 1.0 mM phosphate buffer containing 0.6 mM EDTA, pH 7.4. (D) After lyophilization, the enzyme was dissolved in 0.01 M phosphate buffer and applied to a 2.5 × 30-cm column of DEAE–Sephadex, A-50. Elution was performed with a linear 0- to 0.5-M NaCl gradient in the same buffer, the DDT-ase being eluted at 0.36–0.4 M NaCl (Fig. 1B). The active 1.0-ml fractions were combined, dialyzed, and lyophilized to yield the final pure enzyme for the studies described.

Enzyme assay. Lipke and Kearns (2) described a spectrophotometric assay for DDT-ase involving the use of a suspension of DDT in egg-yolk lipoprotein and its enzymatic conversion to DDE, which absorbs at 260 mµ due to the ethylenic double bond. Despite repeated efforts, we could not reproduce this assay, and a modified assay was, therefore, developed. One tenth milliliter of 1.0 mM DDT dissolved in ethylene glycol and 0.05–0.1 ml of enzyme solution were placed in a small flask and allowed to stand for 5 min at room temperatures to assure enzyme aggregation. 0.2 ml of 0.1 M GSH in 0.2 M phosphate buffer, pH 7.4, and sufficient of the above buffer to make a final volume of 1.0 ml was added, thoroughly mixed, and the absorbance at 260 mµ immediately measured against a blank similar to the above

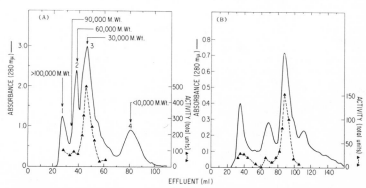

Fig. 1. A. Elution of DDT-ase from Sephadex G-100. Protein (310 mg) in 3.0 ml of 0.01 M phosphate buffer, pH 7.4, containing 0.01 M GSH was applied to a 2.5 × 40-cm column previously equilibrated with the same buffer. One-milliliter fractions were collected at a flow rate of 0.3 ml/min at 4°. The arrows indicate the elution volumes corresponding to the approximate molecular weights calculated from the same column previously calibrated according to Whitaker (29). B. Protein profile obtained by subjecting peak No. 3 (A) to column chromatography on DEAE–Sephadex A-50. Sixty milligrams of protein in 1.0 ml of 0.01 M phosphate buffer, pH 7.4, was applied to a column 2.5 × 30 cm, previously equilibrated with the same buffer. Elution was carried out with 200 ml of a linear gradient of 0–0.5 M NaCl in 0.01 M phosphate buffer pH 7.4. One milliliter fractions were collected at a flow rate of 0.3 ml/min at 4°. The enzyme was eluted at 0.36–0.4 M NaCl.

but lacking GSH. The emulsion was returned to the flask and incubated for 30 min at 37° with constant shaking, after which time a second reading was taken (Fig. 2A and B). The $\Delta \epsilon$ 260 was converted to mμmoles of DDE formed by using the millimolar extinction coefficient of 14.5×10^3 $M^{-1}cm^{-1}$ (2). Specific activity is defined as the mμmoles DDT dehydrochlorinated per mg protein/hr.

Due to the dark color of enzyme extracts during the initial stages of purification, the spectrophotometric assay could be employed only after the Sephadex G-50 purification step. Prior to this stage the enzyme had to be assayed by the more laborious extraction procedure (1).

Analytical procedures. Polyacrylamide gel electrophoresis was performed essentially according to Davis (3) except that polymerization was carried out with riboflavin to avoid the possible deleterious effects of persulfate (4, 5). Protein was measured by the micro Benedict procedure (6). Sulfhydryl groups were determined spectrophotometrically by the PCMB method (7), the mercurial being standardized daily against GSH under conditions identical to those of the assay. The colorimetric nitroprusside method (8), used to distinguish between protein and dithiothreitol thiol groups[3] (9), was slightly modified because the recommended 6.0 M NaCl salted out the DDT-ase. By reducing the NaCl concentration to 1.0 M the enzyme remained in solution at the cost of a 50% reduction in color yield.

Tryptic peptides were prepared by digesting salt-free, heat-denatured DDT-ase with 2.5% by weight of trypsin dissolved in 0.001 N HCl. The pH was adjusted to, and maintained at 8.0 overnight with a Radiometer pH-stat (10). The reaction was terminated by lowering the pH to 2.5, a small amount of insoluble precipitate was centrifuged down, and the clear supernatant fluid was taken to dryness on a rotary evaporator. The dried residue was dissolved in a minimum of distilled water and applied to a sheet of Whatman 3 MM paper for finger-printing. Chromatography in the first direction was performed in butanol-pyridine–acetic acid–H₂O (15:10:3:12), and electrophoresis in the second dimension in pyridine-acetate buffer pH 3.7 (11).

Amino acid composition of the hydrolyzed (12), performic acid oxidized (13) protein was determined on a Beckman "Unichrom" analyzer

<hr>

[3] The rationale for this procedure lies in the fact that dithiothreitol gives a low color yield with nitroprusside, amounting to only 4% of the color given by cysteine (9). At the time of these experiments we were unaware of the better procedure described by Zahler and Cleland (30).

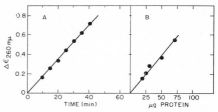

FIG. 2. A. The time course of DDT dehydrochlorination by housefly DDT-ase. The incubation mixture contained: 0.1 ml of 1.0×10^{-3} M DDT in ethylene glycol, 0.05 ml enzyme containing 85 μg protein, 0.2 ml of 0.1 M GSH in 0.2 M phosphate buffer, pH 7.4, and 0.65 ml of the same phosphate buffer. Incubation was performed at 37°, with constant shaking in a Dubnoff water bath except at the times indicated, when the flask contents were transferred to the spectrophotometer cuvette. B. Effect of enzyme concentration on the rate of DDT dehydrochlorination by housefly DDT-ase. Experimental conditions were as described above except that each flask was incubated for 30 min.

according to Spackman *et al.* (14) using cyanide-ninhydrin (15). Cysteine carried through these procedures was recovered to the extent of 90–92% and this value was used in the calculations. Tryptophan was measured spectrophotometrically (16). C-terminal amino acids liberated by carboxypeptidase A were determined on the amino acid analyzer (17). N-terminal residues were isolated as the DNP derivatives and identified on two-dimensional silica gel TLC, using *tert*-amyl alcohol–NH₄OH (4:1) in the first direction and 1 M NaH₂PO₄–0.5 M Na₂HPO₄ in the second direction (18). Estimation of molecular weights by gel filtration and sucrose-gradient centrifugation were performed as described earlier (1).

RESULTS

Effect of reducing agents on DDT-ase aggregation. Gel electrophoresis of DDT-ase purified in the presence of either 0.01 M GSH or mercaptoethanol in the preparative buffers revealed a single protein band of relatively high mobility, providing 0.1 M mercaptoethanol or dithiothreitol is included in both the protein sample and upper buffer (Fig. 3A). This single band remained as such to the point where it could be run off the gel. However, enzyme purified as above, but run on the gel in the absence of

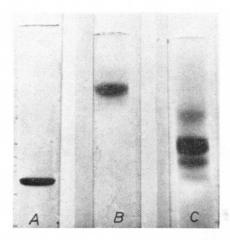

FIG. 3. Disc electrophoresis of housefly DDT dehydrochlorinase. Electrophoresis was performed as described in the text, at 2 mA per tube. Protein was stained with amido schwartz. A, 40 μg enzyme run for 1 hr in the presence of 0.1 M mercaptoethanol in the upper buffer and sample. B, as above, but in the complete absence of reducing agent. C, as B above, but run for 2.5 hr. The protein migrates anodally toward the bottom of the figure.

reducing agent resulted in a single, relatively slowly migrating band for experiments of about 60-min duration (Fig. 3B), and this single band resolved into four if the length of the run was extended to about 150 min (Fig. 3C). Lowering the mercaptoethanol or dithiothreitol concentration in the sample and electrophoresis buffer from 0.1 M to either 0.01 or 0.001 M generally resulted in two protein bands with mobilities between the slower- and faster-running single bands.

It was conjectured on the basis of our earlier work (1) that the single, faster-migrating band was the monomeric form of the enzyme, the slow-moving band was the autoaggregated tetramer produced by the concentrating effect of the upper stacking gel, and the intermediate bands at low reducing agent concentration were the dimer and trimer, respectively. This assumption was verified by the following experiment. The faster- and slower-migrat-

ing single components obtained as described were cut out from unstained gels, the pulverized gel discs were extracted at 4° with 0.2 M phosphate buffer, pH 7.4, overnight, and the protein molecular weights determined by sucrose-gradient centrifugation. The approximate values were 30,000 and 120,000, respectively, and both bands had enzymatic activity. These molecular weights are the same as those found earlier by gel filtration of the enzyme on Sephadex G-200 followed by chromatography on DEAE-Sephadex, and ascribed to the respective monomeric and tetrameric forms of the enzyme (1). Dissociation of the tetrameric enzyme by mercaptoethanol could be directly demonstrated as follows: the tetrameric, unstained gel band obtained by electrophoresis for 45 min in the absence of reducing agent was cut out and introduced with a thin layer of unpolymerized upper stacking gel into a second tube containing preformed lower gel. This tube was then run under identical conditions except for the addition of 0.1 M mercaptoethanol to the upper buffer. After 60 min four protein bands similar to those depicted in Fig. 3C were obtained. The fact that all four oligomers were produced instead of the single monomeric band expected in the presence of mercaptoethanol is undoubtedly due to inadequate concentration of thiol in the gel; mercaptoethanol, being uncharged, enters the gel only by diffusion, and, as noted above, low concentration of the mercaptan is insufficient to maintain the enzyme in the monomeric form. It may be noted that, in our hands, incorporation of the thiol in the gel inhibited photopolymerization with riboflavin as the sole initiator.

Our earlier observations suggested that the mono- and dimeric forms of the protein were enzymatically inactive (1); we now find that dehydrochlorination occurs when these subunits, eluted either from polyacrylamide gels or during gel filtration on Sephadex G-100 (Fig. 1A), are assayed by the present procedure. This discrepancy is explained by the fact that, as observed previously (1), DDT induces aggregation of the enzyme into the active tetrameric form; in the earlier work the DDT level

13

was undoubtedly too low to bring about the rapid and complete association of the monomeric enzyme such as now can be achieved by the 1.0 mM suspension employed in the newer assay. Whether unaggregated monomers *per se* are active, therefore, cannot easily be tested.

Chemical evidence for the identity of DDT-ase monomers. DDT-ase monomers cannot be individually distinguished on the basis of molecular size (gel filtration or sucrose-gradient centrifugation), by charge differences (ionophoresis on cellulose polyacetate strips or ion-exchange chromatography), or by a combination of both of these (polyacrylamide electrophoresis). The following chemical evidence, however, strongly suggests that each of the four monomers probably has the same chemical composition.

The overall amino acid composition of DDT-ase lacks any special distinguishing features (Table I). Of particular interest, however, are the 32 cysteic acid residues to be discussed below, and the 116 lysine and 44 arginine residues. A tryptic hydrolyzate of a protein composed of four chains of unique sequence and containing 160 lysine and arginine residues would be expected to yield a corresponding number, approximately, of tryptic peptides. If, on the other hand, the protein were composed of four identical subunits, only about one quarter of this number of peptides would be anticipated (19). As indicated in Fig. 4, a typical fingerprint pattern of a DDT-ase tryptic digest yielded 43–45 peptides, i.e., about a fourth of the total number of lysine and arginine residues. The fingerprinting experiments, therefore, indicate that DDT-ase is probably composed of four identical chains.

The above hypothesis gains additional support from end-group analyses. Hydrolysis of either native or performic acid oxidized DDT-ase with carboxypeptidase A for 20–60 min resulted in the liberation of several C-terminal amino acids, among which serine and methionine were predominant. In short-term (5-min) incubations these were essentially the only two amino acids liberated. Carboxypeptidase A is stated to release serine relatively more

TABLE I

Amino Acid Composition of Housefly DDT-Dehydrochlorinase Based on a Molecular Weight of 120,000

Amino acid	Residues per mole[a] after hydrolysis		
	24 hr	48 hr	72 hr
Lysine	114	116	118
Histidine	27	30	28
Arginine	41	42	42
Aspartic acid[b]	168	176	173
Threonine	56	54	53
Serine[c]	49	51	43
Glutamic acid[b]	152	150	153
Proline	67	62	60
Glycine	140	138	133
Alanine	102	100	102
Half-cystine[d]	30	32	34
Valine	92	94	90
Methionine	17	16	15
Isoleucine	76	76	76
Leucine	112	114	113
Tyrosine	44	44	44
Phenylalanine	58	59	57
Tryptophan[e]	22		

[a] Three separate enzyme preparations were analyzed in duplicate, one for each time of hydrolysis. Duplicate analyses agreed to within 5%. Values have been rounded off to the nearest whole number.

[b] Includes the corresponding amides.

[c] Value obtained by extrapolation to zero-time hydrolysis.

[d] Twenty-four-hour air-oxidized native sample; 48 and 72 hr, estimated as cysteic acid after performic acid oxidation (14).

[e] Determined by spectrophotometry (16).

slowly than methionine (17), while in our experiments the amount and rate of serine liberation from DDT-ase always exceeded that of methionine. Further, with DDT-ase unfolded in 0.1% (w/v) SDS, the amount of serine released after 40-min incubation was 3.87 μmoles/μmole DDT-ase tetramer, or a yield of 97% assuming four chains each with a C-terminal serine. For this particular preparation, 2.5 μmoles methionine were also released. We conclude, therefore, that serine is the sole C-terminus with methionine being the penultimate residue.

Turning next to the N-terminal end, FDNB-treated tetrameric DDT-ase after hydrolysis revealed only a single DNP-

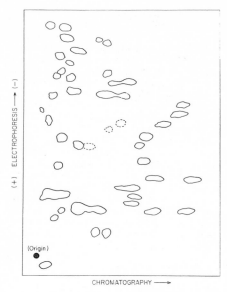

FIG. 4. Tracing of a typical fingerprint pattern obtained from a tryptic digest of DDT-ase. Spots represent staining of peptide with ninhydrin. Dotted areas represent very weak ninhydrin-reacting spots. For experimental details, see text.

amino acid on two-dimensional TLC; the position of this spot corresponded to either leucine or isoleucine, which could not be satisfactorily differentiated even in several solvent systems. Attempts to regenerate the original amino acid from the DNP derivative in the ether phase by alkaline hydrolysis (20) were unsuccessful. These experiments demonstrate, nevertheless, that DDT-ase probably possesses only a single leucine or isoleucine N-terminal residue—a finding again consistent with the postulate of four identical chains.

Sulfhydryl groups of DDT-ase. As indicated in Table I, performic acid-oxidized DDT-ase contains 32 moles cysteic acid/mole tetrameric enzyme. Amino acid analysis of the native enzyme performed under conditions such that all cysteine residues were air-oxidized to cystine (12), yielded 32 half-cystine residues/mole enzyme. Because these methods do not distinguish

between cysteine and cystine residues, further experiments were directed toward elucidation of the number of -SH groups, and possible presence of disulfide bonds in the protein.

A typical PCMB titration curve, of DDT-ase unfolded in freshly prepared 8 M urea, is shown in Fig. 5. Reaction was rapid throughout the course of the titration, and revealed that 32 μmoles PCMB reacted/μmole tetrameric enzyme. These results, together with the amino acid analyses, indicate that all the protein thiol groups are attributable to cysteine residues. Additional evidence that disulfide bonds are lacking was obtained by use of dithiothreitol, which quantitatively reduces disulfides (9). It was reasoned that if such bonds were present in DDT-ase, treatment with dithiothreitol would reduce them to sulfhydryls reactive with nitroprusside. However, no sulfhydryls could be detected with the nitroprusside reagent after reacting the DDT-aggregated enzyme with a 1000-fold excess of dithiothreitol overnight; the amount of color was no greater than that

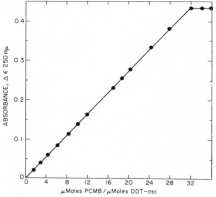

FIG. 5. The titration of DDT-ase with PCMB in the presence of 8 M urea. DDT-ase (2×10^{-6} M) was dissolved in 1.0 ml of 8 M urea in 0.05 M Tris-HCl buffer, pH 7.0. Aliquots (0.02 ml) of a freshly assayed PCMB solution (0.36×10^{-3} M) were added to the enzyme solution, and the change in absorption at 250 mμ was measured. Appropriate corrections for blanks and volume changes have been made in measuring the absorbancy.

15

obtained with the same amount of dithio-threitol alone. When the enzyme was sub-sequently unfolded in 8 M urea, the nitro-prusside color corresponded to 30 μmoles of thiol/μmole enzyme. These results, therefore, indicate that the four monomers of tetrameric DDT-ase are not joined by interchain disulfide bridges.

Availability of thiol groups. Titration of the sulfhydryl groups of native DDT-ase in pH 7.0 Tris–HCl buffer with small in-crements of PCMB is a slow reaction, requiring several hours to attain completion (Fig. 6). The addition of 1.14 μmoles PCMB/enzyme thiol group required 8 hr for complete reaction, but when this ratio was increased to 7, the reaction was com-pleted in 10 min. These results suggest that few, if any, of the enzyme thiol groups are freely exposed on the protein surface. Furthermore, the form of the PCMB titra-tion curve [cf(21)] suggests the existence of two categories of sulfhydryl groups: a faster-reacting set that has completely reacted after *ca.* 1 hr, and a slower-reacting class which titrates at an almost constant rate for the duration of the experiment. The zero time intercept of the extrapolated linear portion of the curve gives the number of faster reacting thiols, i.e., 6, to the nearest whole integer, and hence the slower-reacting class contains 26 thiols/mole. The rate of reaction of this latter class is given by the slope of the straight line, and the rate of the faster-reacting set by the dif-ference between the overall curve and the extrapolated line, from which it appears that the faster-reacting sulfhydryls titrate about 1.3 times more rapidly than the slower ones. Presumably, the more reactive thiols are either in a less hydrophobic en-vironment, or unfolding of the protein

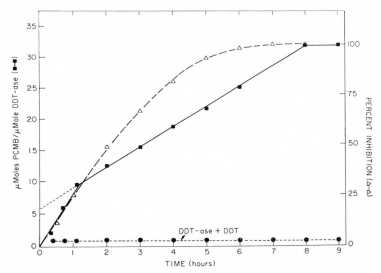

Fig. 6. The reaction of PCMB with DDT-ase. PCMB (7.0×10^{-1} μmoles) were added to 1.6×10^{-2} μmoles DDT-ase (1.4 μmoles PCMB per enzyme thiol) in a final volume of 3.0 ml of 0.05 M Tris–HCl buffer, pH 7.0. Absorbance changes at 250 mμ were measured at the times indicated, and 0.1-ml aliquots of the incubation mixture were periodically removed for assay of enzyme activity. To examine the effect of DDT, 2.3×10^{-3} μmoles enzyme in 0.9 ml of 0.05 M Tris–HCl buffer, pH 7.0, were preincubated for 3 hr with 0.1 μmole DDT in 0.1 ml ethylene glycol. Subsequently, 5.2×10^{-1} μmoles PCMB were added, and the Δ250 measured against a blank lacking enzyme.

with concomitant exposure of previously inaccessible thiols occurs in two phases.

Inhibition studies also indicated two categories of sulfhydryl groups; the extent of DDT-ase inhibition by PCMB was directly proportional to the number of sulfhydryls modified for the first 10 thiols, but thereafter enzyme inhibition was disproportionately greater than the number of thiols titrated until 32 groups had reacted, at which point the enzyme was completely inactive (Fig. 6). It is especially noteworthy that the substrate, DDT, completely protects the fully aggregated enzyme against PCMB inhibition. No free sulfhydryls could be detected even with 400 moles excess of PCMB nor, as noted above, with nitroprusside, after the enzyme had been preincubated with DDT. Since the expected number of sulfhydryls could be determined with either reagent when the aggregated enzyme was subsequently unfolded in 8 M urea, the masking and unmasking of the protein thiols is clearly a reversible reaction.

DDT-ase unfolded in 8 M urea or 0.1% SDS is completely inactive. Attempts to reactivate the urea-treated enzyme by passage through a Sephadex G-15 column to remove urea were unsuccessful. Neither could we titrate the thiol groups of detergent-treated enzyme with PCMB; upon addition of the mercurial the solution immediately became cloudy, and no measurements were possible.

The inhibitory effect of PCMB on DDT-ase activity was only partially reversible with either GSH or mercaptoethanol, but the results were anomalous. Thus, in typical experiments, enzyme was completely inhibited by preincubation with 200 μmoles PCMB/μmole enzyme for 30 min at room temperature. A 15-fold moles excess of GSH, or 25-fold excess of mercaptoethanol was added, and aliquots were then assayed by the routine procedure during the ensuing 4 hr. In the case of GSH, no activity was demonstrable after 40 min, but after 60 min 44% of the original activity was regained with no subsequent change thereafter. No activity was detectable 90 min after addition of the mercaptoethanol,

while after 120 min 32% reactivation occurred, again with no subsequent change.

Aliphatic alcohols are reported to weaken the hydrophobic conformational bonds of some proteins, and to expose thereby certain previously unreactive thiol groups (22, 23). To examine whether DDT-ase sulfhydryl groups could be similarly exposed, a small excess of PCMB was added to the enzyme in the presence of 4.5% (v/v) methanol, ethanol and n-propanol, and 2.3% (v/v) n-butanol, and enzyme activity was measured as a function of time. It may be noted that, in controls, no inhibition was observed with the alcohols alone, and that, within the range of inhibition experimentally observed, enzyme inactivation is proportional to the number of sulfhydryl groups modified. As shown in Fig. 7, as the alcohol becomes progressively more hydrophobic, so the rate of enzyme inactivation increases. This finding strongly suggests that conformational changes induced by the alcohols expose thiol groups previously buried in, or masked by, hydrophobic portions of the protein fabric. Once these are exposed, reaction with PCMB is facilitated.

Lipoprotein nature of DDT-ase. During the course of the above experiments, a number of observations indicated that DDT-ase was a lipoprotein. The evidence may be summarized as follows. A drop of the concentrated, purified enzyme when dried on

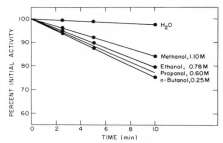

Fig. 7. The effect of aliphatic alcohols on the rate of inactivation of DDT-ase with PCMB. DDT-ase (1.1×10^{-6} M) in 8.5×10^{-2} M phosphate buffer, pH 7.4, was allowed to react at 27° with 5.0×10^{-5} M PCMB in each of the alcohols indicated. Aliquots of the reaction mixture were periodically removed and assayed as described.

17

filter paper appears translucent to transmitted light and gives a positive lipid stain with Sudan black B, and a positive colorimetric reaction for phosphorus. Extraction of the protein by refluxing with chloroform–ethanol 2:1 (v/v) yielded an oily, water-insoluble material after evaporating off the solvent. This lipid, after ashing with concentrated H_2SO_4, contained 6 μmoles P/μmole DDT-ase, which suggests that the lipid moiety is probably a phospholipid.

DISCUSSION

The present experiments, in conjunction with those presented earlier (1), leave little doubt that enzymatically active DDT dehydrochlorinase from the P_2/sel strain of DDT-resistant houseflies is a tetrameric protein. Reducing agents, such as β-mercaptoethanol or dithiothreitol, promote dissociation of the tetramer, and the addition of these reagents or of GSH to the preparative solutions results in isolation of the enzyme in its monomeric state. The effect of GSH is unique: first, it is an essential co-factor in the enzymatic reaction, and cannot be replaced by any of a wide variety of -SH compounds with the exception of cysteinyl glycine, which is less effective (2). Second, we have observed that GSH, unlike β-mercaptoethanol or dithiothreitol, actually stabilizes the tetrameric protein in its aggregated state. Whether this apparently contradictory effect of GSH has anything to do with its role in enzymatic dehydrochlorination can only be conjectured.

Although the possibility of minor variations in subunit amino acid composition cannot be excluded, our amino acid, fingerprinting and end-group analyses strongly suggest that the four DDT-ase monomers have an identical amino acid sequence, and our failure to distinguish between them by physical means supports this notion. Goodchild and Smith (24) subjected crude or semipurified DDT-ase from DDT-resistant houseflies to isoelectric focusing, and obtained four peaks of activity with runs of 3 days' duration, but only a single, larger peak when the length of the run was extended to 9 days. Their observations are in a sense the converse of ours on gel electrophoresis, where the single band of pure tetrameric enzyme, in the absence of reducing agents, resolves into the monomer, dimer, trimer, and tetramer during runs of longer duration. It may be speculated that Goodchild and Smith's observations are also a manifestation of oligomer formation, the differences in pI being due to variations in net charge associated with progressive interchain bonding, e.g., to release or binding of ions (25), masking or exposure of charged groups due to conformational change, or the like.

The interchain bonds of DDT-ase oligomers appear to be weak and readily reversible. Thus, monomers, in addition to promptly aggregating into tetramers in the presence of substrate (DDT), will also do so spontaneously, albeit much more slowly, when in relatively high concentration at low ionic strength. Conversely, in high salt and at low protein concentrations in the absence of GSH the tetramers tend to dissociate (unpublished observations). Such reactions are indicative of loose bonding, and are compatible with the demonstrated absence of interchain disulfide bridges.

Housefly DDT-ase with 32 cysteines per 120,000 mol wt can now be added to the extensive list (26) of "-SH enzymes." It seems unlikely, however, that cysteine is associated with the enzyme active center. As in the case of fumarase (23) and certain other enzymes, the thiol groups of the native protein react very sluggishly with PCMB unless the latter is added in large molar excess, or unless the protein is first unfolded in urea. These results suggest that the thiols are probably not on the protein surface, and are unavailable for reaction with the mercurial either because they are buried in the protein matrix, or are enclosed in a hydrophobic environment. The finding that a series of aliphatic alcohols of increasing chain length (and, therefore, increasing hydrophobicity) is progressively more effective in exposing enzyme sulfhydryls is indicative of the hydrophobic character of this environment.

The thiol groups of many native enzymes are not equally reactive to solvent mer-

18

curials or alkylating agents. In the case of DDT-ase, 6 thiols react relatively more rapidly with PCMB than the remaining 26. Progressive titration with mercurial results in a proportional loss of enzyme activity for the first 10 -SH groups titrated but a disproportionately greater inhibition thereafter until all 32 -SH groups have been titrated when the enzyme is completely inactive. These findings, together with the anomalous and incomplete reversal of the PCMB-inhibited enzyme on the addition of excess GSH, suggest that such inhibition is not due to a specific reaction with thiols, but rather to generalized and only partially reversible change in protein conformation.

Conformational changes must also be responsible for the striking effect of DDT in masking all 32 protein thiols. The fact that substrate protects certain enzymes against PCMB inhibition is well documented (26), and cases are known (e.g., 27) where no thiol groups can be titrated unless the enzyme is first disaggregated into its component subunits. Even the eight thiols of each native DDT-ase monomer are only slowly titratable with PCMB, and when these subunits aggregate to form the active enzyme the final conformation is such that no thiol is any longer exposed to the solvent.

The solubility of DDT in water is only 3×10^{-9} moles/1 (28); for all practical purposes it is insoluble in aqueous solvents. In view of our evidence that DDT-ase may be a phospholipoprotein, and the known solubility of DDT in lipids, it seems plausible to suggest that the DDT first "dissolves" in the lipid moiety of the enzyme prior to subsequent dehydrochlorination. We have not yet succeeded in delipidating the protein under nondenaturing conditions, although the phospholipid moiety *per se* is now under investigation. It is hoped that experiments along these lines may shed some light on the essential participation of GSH in the enzymatic reaction.

ACKNOWLEDGMENTS

This study was supported by Grants TW00239-03 from the Division of Research Grants and Fellowships, U. S. Public Health Service, and 2.107.4314 (66-49) of the Comisión Científica, Facultad de Medicina, Universidad de Chile.

REFERENCES

1. Dinamarca, M. L., Saavedra, I., and Valdés, E., *Comp. Biochem. Physiol.* **31**, 269 (1969).
2. Lipke, H., and Kearns, C. W., *J. Biol. Chem.* **234**, 2129 (1959).
3. Davis, B. J., *Ann. N. Y. Acad. Sci.* **121**, 404 (1964).
4. Brewer, J. M., *Science* **156**, 256 (1967).
5. Fantes, K. H., and Furminger, I. G. S., *Nature London* **215**, 750 (1967).
6. Goa, J., *Scand. J. Clin. Lab. Invest.* **5**, 218 (1953).
7. Boyer, P. D., *J. Amer. Chem. Soc.* **76**, 4331 (1954).
8. Grunert, R. R., and Phillips, P. H., *Arch. Biochem.* **30**, 217 (1951).
9. Cleland, W. W., *Biochemistry* **3**, 480 (1964).
10. Smyth, D. G., *Methods Enzymol.* **11**, 214 (1967).
11. Katz, A. M., Dreyer, W. J., and Anfinsen, C. B., *J. Biol. Chem.* **234**, 2897 (1959).
12. Moore, S., and Stein, W. H., *Methods Enzymol.* **6**, 819 (1963).
13. Hirs, C. H. W., *Methods Enzymol.* **11**, 197 (1967).
14. Spackman, D. H., Stein, W. H., and Moore, S., *Anal. Chem.* **30**, 1190 (1958).
15. Rosen, H., Berard, C. W., and Levenson, S. M., *Anal. Biochem.* **4**, 213 (1962).
16. Bencze, W. L., and Schmid, K., *Anal. Chem.* **29**, 1193 (1957).
17. Ambler, P., *Methods Enzymol.* **11**, 155 (1967).
18. Fraenkel-Conrat, H., Harris, J. I., and Levy, A. L., *Methods Biochem. Anal.* **11**, 359 (1955).
19. Kanarek, L., Marler, E., Bradshaw, R. A., Fellows, R. E., and Hill, R. L., *J. Biol. Chem.* **239**, 4207 (1964).
20. Lowther, A. G., *Nature London* **167**, 767 (1951).
21. Gold, A. M., *Biochemistry* **7**, 2106 (1968).
22. Cecil, R., and Thomas, M. A. W., *Nature London* **206**, 1317 (1965).

19

23. ROBINSON, G. W., BRADSHAW, R. A., KANA-
    REK, L., AND HILL, R. L., *J. Biol. Chem.*
    242, 2709 (1967).
24. GOODCHILD, B., AND SMITH, J. N., *Biochem.
    J.* **117,** 1005 (1970).
25. TANFORD, C., "Physical Chemistry of Macro-
    molecules," Wiley, New York, 1961.
26. WEBB, J. L., "Enzyme and Metabolic Inhibi-
    tors," Vol. **11,** p. 729. Academic Press, New
    York, 1966.
27. SHAPIRO, B. M., AND STADTMAN, E. R., *J.
    Biol. Chem.* **242,** 5069 (1967).
28. BOWMAN, M. C., ACREE, F., JR., AND CORBETT,
    M. K., *J. Agr. Food Chem.* **8,** 406 (1960).
29. WHITAKER, J. R., *Anal. Chem.* **35,** 1950 (1963).
30. ZAHLER, W. L., AND CLELAND, W. W., *J. Biol.
    Chem.* **243,** 716 (1968).

Induction by DDT and Dieldrin of Insecticide Metabolism by House Fly[1] Enzymes[2,3]

Frederick W. Plapp, Jr. and John Casida

Exposure of mammals to chlorinated insecticides such as DDT or dieldrin induces increased activity of liver microsomal mixed-function oxidases in mammals (Hart and Fouts 1963, Morello 1965). The same or similar enzyme systems occur in insects (Arias and Terriere 1962) but it is not known if they are generally inducible in terms of insecticide metabolism or what role induction may play in the development of resistance. If microsomal induction does occur in insects, it may lead to an adaptive mechanism for the development of resistance to either the inducing insecticide or to other insecticides detoxified by microsomal enzymes, a possibility first suggested by Agosin (1963). This is a logical consequence of the use of insecticides because insects are exposed frequently to sublethal concentrations of one or more insecticides during the period in which residues of insecticides dissipate from treated areas.

House fly, *Musca domestica* L., mixed-function NADPH-dependent oxidases metabolize a great variety of insecticides (Plapp and Casida 1969) and high levels of these oxidases frequently occur in resistant strains (Schonbrod et al. 1968, Tsukamoto et al. 1968). Since induction, intentional or otherwise, may increase the level of detoxifying enzymes, it is important to determine the extent to which microsomal enzyme systems can be induced in insects. It is known that DDT induces RNA and protein synthesis in the house fly (Balazs and Agosin 1968, Ishaaya and Chefurka 1968) and that DDT or phenobarbital in-

[1] Diptera: Muscidae.
[2] This study was supported in part by grants from the U. S. Public Health Service, National Institutes of Health (Grants GM-12248 and ES-00049) and the U. S. Atomic Energy Commission (Contract AT(11-1)-34, Project Agreement no. 113.

[3] The authors thank Judith Engel, Irene Jackson, Louis Lykken, and Loretta Gaughan for their invaluable assistance.

21

duce DDT metabolism in house flies and *Triatoma infestans* (Klug) (Gil et al. 1968, Agosin et al. 1969). However, other investigators have failed to demonstrate the induction of mixed-function oxidase activity in insects (Chakraborty and Smith 1967, Meksongsee et al. 1967, Oppenoorth and Houx 1968).

In the present report, we demonstrate that both DDT and dieldrin are effective inducers of the metabolism of several insecticides by homogenates of abdomens prepared from house flies resistant to chlorinated insecticides. The induction was achieved by feeding massive dosages of DDT or dieldrin to flies for several days prior to preparation of homogenates for in vitro metabolic experiments.

MATERIALS AND METHODS.—*House Flies.*—Two strains of *M. domestica* resistant to chlorinated insecticide chemicals were used. One was the Orlando DDT strain, immune to DDT and resistant to dieldrin and other cyclodiene insecticides (Hoyer and Plapp 1966). The other was a dieldrin-resistant, but DDT-susceptible strain carrying the visible recessive mutant *curly wing* (Hoyer and Plapp, unpublished work, USDA, Corvallis, Ore. 1967). Both strains have the low microsomal oxidase level characteristic of susceptible strains and strains in which resistance depends on nonoxidative mechanisms (Schonbrod et al. 1968).

Treatment.—Flies were fed either a 1:1 diet of sucrose and powdered whole milk or the same diet containing 1000 ppm technical DDT or 100 ppm 90% dieldrin. Treated diets were prepared by adding required amounts of insecticide in 10 ml acetone to 100 g diet and manually mixing until the solvent evaporated. Acetone was added similarly to the control diet. Preliminary tests established that these levels of insecticide were effective as inducers of enzymes which metabolize insecticides and that at least a 4-day feeding period was necessary for the demonstration of inductive effects.

ASSAY.—The in vitro NADPH-dependent oxidative metabolism assays were made by previously described methods (Tsukamoto et al. 1968). Substrates were ^{14}C-labeled samples of aldrin, allethrin, propoxur, DDT, and diazinon. The sources and sites of labeling of the substrates as well as the TLC methods used to determine metabolism have been given previously (Plapp and Casida 1969). Enzyme preparations consisted of homogenates of house fly abdomens (equal numbers of each sex) prepared in phosphate-sucrose buffer containing bovine serum albumen. Concentration was 3 fly abdomens/assay with all substrates except allethrin where 1 abdomen/assay was used. All flies for each assay with each strain were reared and tested simultaneously. The results from

22

Table 1.—Induction of NADPH-dependent metabolism of certain insecticides by homogenates of abdomens of house flies fed high dietary levels of **DDT** or dieldrin.

| Substrate | Concentration of substrate (mμmoles) | Percent of substrate metabolized in 30 min (mean + SE) | | | | |
| | | Orlando DDT-resistant strain | | | "Curly wing" dieldrin-resistant strain | |
		Control diet	1000 ppm DDT in diet	100 ppm dieldrin in diet	Control diet	100 ppm dieldrin in diet
Aldrin	0.14	37±5	54±1	71±5	31±4	46±2
Allethrin	1	40±9	57±2	77±1	39±3	68±3
Propoxur	3	14±3	17±4	21±3	12±2	18±3
DDT	8.8	1.8±1.2	3.4±1.2	2.5±1.1	1.3±1.2	1.8±0.5
Diazinon	3	11±1	16±2	17±2	9±3	23±5

3 replicate experiments were averaged.

RESULTS.—The data (Table 1) show that both DDT and dieldrin are effective inducers of insecticide metabolism in the house fly. The effect is apparently nonspecific as the rate of metabolism of all substrates tested was increased by both treatments. The increase in metabolism varied from approximately 20% with propoxur for flies fed DDT and 35% with DDT for flies fed dieldrin to as much as 150% for diazinon for flies fed dieldrin. In general, the increases ranged from 50 to 100%.

With the Orlando DDT strain, a direct comparison was possible of the relative effectiveness of DDT and dieldrin as inducers. In general, dieldrin was more effective than DDT. Not only was the dietary level of dieldrin necessary to produce inductive effects lower, but the average increase in metabolism was greater. This was verified by other experiments (results not shown) in which DDT at 100 ppm was less active as an inducer than DDT at 1000 ppm, and dieldrin at 1000 ppm was no more active than dieldrin at 100 ppm.

DISCUSSION.—The data demonstrated that exposure of house flies to chlorinated insecticides enhances metabolism, not only of the inducing chemicals, but of other types of insecticides as well. Thus, a possible consequence of exposure of house fly populations to chlorinated insecticides is an increase in ability to degrade organophosphates, carbamates, and pyrethroids. This phenomenon may explain in part why resistance to these latter types of insecticides has developed so rapidly in recent years.

It should be noted that the conditions of exposure used to demonstrate induction in these experiments may not be comparable to normal conditions of use of either DDT or dieldrin; the high dosages of insecticides used as inducers and the constant exposure for a period of days seem unlikely to occur under field conditions.

REFERENCES CITED

Agosin, M. 1963. Present status of biochemical research on the insecticide resistance problem. Bull. World Health Organ. 29: Suppl. 69–76.

Agosin, M., N. Scaramelli, L. Gil, and M. E. Letelier. 1969. Some properties of the microsomal system metabolizing DDT in *Triatoma infestans*. Comp. Biochem. Physiol. 29: 785–93.

Arias, R. O., and L. C. Terriere. 1962. The hydroxylation of naphthalene-C^{14} by house fly microsomes. J. Econ. Entomol. 55: 925–9.

Balazs, I., and M. Agosin. 1968. The effect of 1,1,1-trichloro-2,2-bis (*p*-chlorophenyl) ethane on ribonucleic acid metabolism in *Musca domestica* L. Biochem. Biophys. Acta 157: 1–7.

Chakraborty, J., and J. N. Smith. 1967. Enzymic oxidation of some alkylbenzenes in insects and vertebrates. Biochem. J. 102: 498–503.

Gil, L., B. C. Fine, M. L. Dinamarca, I. Balazs, J. R. Busvine, and M. Agosin. 1968. Biochemical studies on insecticide resistance in *Musca domestica.* Entomol. Exp. Appl. 11: 15–28.

Hart, L. G., and J. Fouts. 1963. Effects of acute and chronic DDT administration on hepatic microsomal drug metabolism in the rat. Proc. Soc. Exp. Biol. Med. 114: 388–92.

Hoyer, R. F., and F. W. Plapp, Jr. 1966. A gross genetic analysis of two DDT-resistant house fly strains. J. Econ. Entomol. 59: 495–501.

Ishaaya, I., and W. Chefurka. 1968. Effect of DDT on microsomal RNA and protein biosynthesis in susceptible and DDT-resistant houseflies Riv. Parassitol. 29: 289–96.

Meksongsee, B., R. S. Yang, and F. E. Guthrie. 1967. Effect of inhibitors and inducers of microsomal enzymes on the toxicity of carbamate insecticides to mice and insects. J. Econ. Entomol. 60: 1469–71.

Morello, A. 1965. Induction of DDT-metabolizing en zymes in microsomes of rat liver after administration of DDT. Can. J. Biochem. 43: 1289–93.

Oppenoorth, F. J., and N. W. H. Houx. 1968. DDT resistance in the house fly caused by microsomal degradation. Entomol. Exp. App. 11: 81–93.

Plapp, F. W. Jr., and J. E. Casida. 1969. Genetic control of house fly NADPH-dependent oxidases: Relation to insecticide chemical metabolism and resistance. J. Econ. Entomol. 62: 1174–9.

Schonbrod, R. D., M. A. Q. Khan, L. C. Terriere, and F. W. Plapp, Jr. 1968. Microsomal oxidases in the house fly: A survey of fourteen strains. Life Sci. 7 (1) : 681–8.

Tsukamoto, M., S. P. Srivastava, and J. E. Casida. 1968. Biochemical genetics of house fly resistance to carbamate insecticide chemicals. J. Econ. Entomol. 61: 50–55.

Some Factors Affecting Degradation of Organochlorine Pesticides by Bacteria

B. E. LANGLOIS, J. A. COLLINS,[1] and K. G. SIDES
Department of Animal Sciences, Food Science Section
University of Kentucky, Lexington 40506

dichloro-2,2-*bis* (*p*-chlorophenyl) ethane) and smaller amounts of DDE (1,1-dichloro-2,2-*bis*(*p*-chlorophenyl)ethylene). Wedemeyer (16) found the pathway for degradation of DDT by *Aerobacter aerogenes* to be: DDT ⟶ DDD ⟶ DDMU (1-chloro-2,2-*bis*(*p*-chlorophenyl)ethylene) ⟶ DDMS (1-chloro-2,2-*bis*(*p*-chlorophenyl)ethane) ⟶ DDNU (unsym-*bis*(*p*-chlorophenyl)ethylene) ⟶ DDOH (2,2-*bis* (*p*-chlorophenyl)ethanol ⟶ DDA (2,2-*bis* (*p*-chlorophenyl)acetic acid ⟶ DBP (4,4′-Dichlorobenzophenone) or DDT ⟶ DDE. This pathway is similar to the one postulated for the degradation of DDT by the rat (13).

Degradation of other organochlorine pesticides has not been studied in detail. Several workers have reported the degradation of dieldrin (1,2,3,4,10,10-hexachloro-6-6-*epoxy*-1,4, 4a,5,6,7,8,8a-octahydro-1,4-*endo*- *exo*-5,8-dimethanonaphthalene) to aldrin diol (6,7-*trans*-dihydroxydihydroaldrin) (12,16).

Since milk had been found to contain varying amounts of organochlorine pesticides it seemed desirable to obtain information on the extent to which microbial degradation of DDT, dieldrin and heptachlor (1,4,5,6,7,8,8-heptachloro-3a,4, 7,7a-tetrahydro-4-7-methanoindene) does occur in milk.

Introduction

Recent interest has focused on the mechanisms by which microorganisms degrade organochlorine pesticides to less harmful products, since many of them are not biodegradable and may persist for a long time in the soil. Most of the previous work has been with DDT (1,1,1-trichloro-2,2-*bis* (*p*-chlorophenyl)ethane) (2,3,5,7,9,10,13, 14,15). Depending on the animal or microbial species as well as availability of oxygen, as many as eight products have been identified from the degradation of DDT (2,14,15). The first products formed appear to be DDD (1,1-

[1] Present address, Department of Animal Sciences, University of Arkansas, Fayetteville 72071.

Materials and Methods

Organisms. Strains of *Bacillus cereus, Bacillus coagulans, Bacillus subtilis, Escherichia coli, Enterobacter aerogenes, Pseudomonas fluorescens* and *Staphylococcus aureus* were obtained from our collection in the department of animal sciences, food science section.

Culture procedure. The bacteria were routinely grown in trypticase soy broth at the optimum temperature for the species. Working cultures were transferred weekly, whereas the stock cultures were maintained on nutrient agar slants stored at 7 C and transferred at 4-month intervals. The organisms were grown in trypticase soy broth for three successive daily transfers, before inoculation of a 15- to 18-hour culture in flasks containing the test media and pesticide. Flasks containing only pesticide and media were run as controls. All flasks were incubated aerobically at optimum temperature for up to 30 days. Flasks containing *E. coli* and *E. aerogenes* also were incubated anaerobically in a BBL Gas Pak anaerobic jar.

Test media. Growth media to determine the ability of the test bacteria to degrade organochlorine pesticides were: Skimmilk (Difco), Matrix (Galloway-West), and trypticase soy broth (BBL). In addition, lactose, whole casein and α-, β- and γ-caseins were added individually to trypticase soy broth to give concentrations normally found in skimmilk. All media were dispensed into screwcap Erlenmeyer flasks in amounts of from 50 to 300 ml.

Preparation of whole casein and casein fractions. Whole casein was prepared from fresh skimmilk by acid precipitation (8). The α-, β-, and γ-caseins were prepared from whole casein by the urea fractionation method of Hipp et al (8). Purity of whole casein and the casein fractions was determined by thin layer electrophoresis, and the products were stored at 5 C until used.

Pesticides. Degradation of both technical and purified grades of the following pesticides in various media was determined: DDT, dieldrin and heptachlor. Stock solutions of each pesticide were made to contain 15 mg per milliliter in alcohol. The stock solution was diluted with alcohol so that final concentrations of 1 to 200 μg per milliliter of medium could be obtained by adding 2.0 ml or less to flasks containing 100 to 300 ml of sterile medium. Identification of DDT degradation products was by the same procedure as that of Wedemeyer (15). The DDT, DDD, DDE, DBP and DDA standards used in this study are commercially available in high purity. The remainder of the metabolites were synthesized according to the procedures of Peterson and Robison (13).

Extraction of residues and metabolites. After incubation the samples were extracted using both the florisil column cleanup method of Langlois et al (11) and the method of Wedemeyer (15), except that residues were redissolved in hexane rather than in acetonitrile. In addition, some of the skimmilk samples were extracted by the Soxhlet method of Peterson and Robison (13). Extracted samples analyzed by paper chromatography were treated by the acid cleanup procedure of Peterson and Robison (13) to remove interfering substances.

Analysis for pesticide residues and metabolites. Extracted solutions were assayed for residues and metabolites by electron capture gas chromatography, thin layer chromatography and paper chromatography. A Perkin-Elmer Model 811 gas chromatograph with a tritium electron capture detector was used. Borosilicate glass columns (20 mm od by 600 mm) packed with these materials were used for assay: a) 5% DC-11 on 60/80 mesh Gas Chrom Q with column at 190 C and nitrogen flow of 60 ml per minute; b) 10% DC-200 on 100/120 mesh Gas Chrom Q with column at 190 C and nitrogen flow at 100 ml per minute; c) 11% (OV-17 + QF-1) on 80/100 mesh Gas Chrom Q with column at 190 C and nitrogen flow of 120 ml per minute.

Eastman chromagram sheets and developing apparatus were used for thin layer chromatography (1). The method of Mills (4) and the solvent systems of Wedemeyer (15) were used for paper chromatography.

Results and Discussion

Degradation of dieldrin. None of the species was able to degrade dieldrin. Our results agree with those of Chacko et al (5). However, several investigators have reported microbial degradation of dieldrin (12,16). Differences in results might be due to variations in methodology or to differences in strain of species. Two of the species reported to be capable of degrading dieldrin were isolated from dieldrin-treated soil, and may have become dieldrin-tolerant (12); whereas species used in our study were not subjected to such adaptation. Wedemeyer (16) isolated dieldrin metabolites from sonically disrupted cells of *Aerobacter aerogenes*, whereas liquid medium containing whole cells was analyzed in our study. These differences suggest that degradation of dieldrin is intracellular and metabolites are released only by lysing the cell.

Degradation of heptachlor. As with dieldrin none of the bacteria studied was able to degrade heptachlor. In addition, compounds making up some of the impurities in 73% heptachlor also were nondegradable by the bacteria studied. These impurities were heptachlor epoxide, chlordane and gamma chlordane.

Degradation of DDT. Results from degradation of DDT by whole cells of test bacteria in four media under aerobic and anaerobic incubation are in Table 1. Evidently not all species are capable of degrading DDT, amount of degradation is affected by growth medium and amount of degradation is affected by availability of oxygen during incubation.

Generally DDD and DDE were detected in various amounts after two days of incubation with maximum levels being obtained within seven days. At least 30 days of incubation were required to detect all of the metabolites listed in the table.

Whole cells of *P. fluorescens* and *S. aureus*

TABLE 1. Degradation of DDT in various media under aerobic and anaerobic incubation.

Species	Medium[a]	Incubation	Metabolites detected after incubation[b]								
			DDT	DDE	DDD	DDMU	DDMS	DDNU	DDOH	DDA	DEP
Escherichia coli	TSB	Aerobic	A[c]	B	A	C	C	C	C	C	C
	TSB	Anaerobic	B	B	A						
	SM	Aerobic	A		C						
	SM	Anaerobic	A		C						
	TSBC	Aerobic	A		C						
	TSBL	Aerobic	A	C	A						
Enterobacteria aerogenes	TSB	Aerobic	A	B	A	C	C	C	C	C	C
	TSB	Anaerobic	B	B	A						
	SM	Aerobic	A		C						
	SM	Anaerobic	A		C						
Bacillus cereus	TSB	Aerobic	B	B	A	C	C	C	C	C	C
Bacillus coagulans	SM	Aerobic	A		C						
Bacillus subtilis											
Pseudomonas fluorescens	TSB	Aerobic	A								
Staphylococcus aureus	TSB	Aerobic	A								

[a] TSB, trypticase soy broth; SM, skimmilk; TSBC, trypticase soy broth + 3% whole casein; TSBL, trypticase soy broth + 4.5% lactose.
[b] Products detected by one or all of following: electron capture gas chromatography, thin-layer chromatography and paper chromatography.
[c] A, major or only product; B, minor product; C, trace product.

were not capable of degrading DDT, to any great extent, during aerobic incubation in TSB for up to 14 days. Wedemeyer (15) found DDT, DDE, DDD, DDMU and DDNU after DDT was incubated anaerobically with cell-free extracts of *P. fluorescens*. He found less conversion of DDT under aerobic conditions. Whole cells of *P. fluorescens* are unable to degrade DDT aerobically but cell-free extracts are able to convert DDT under both aerobic and anaerobic conditions. Apparently a permeability phenomenon exists with whole cells.

Results with whole cells of *E. coli* and *E. aerogenes* agree with those of other workers (10,15). Wedemeyer (15) found that cell-free extracts of *A. aerogenes* degraded DDT anaerobically to DDD, DDE, DDMU, DDNU, but only to DDD and DDE under aerobic conditions. In our study, anaerobic incubation was for 30 days, and DDMS, DDA and DBP were identified in addition to DDD, DDE, DDMU and DDNU. Generally only DDE and DDD were found after 7 days of incubation.

The *Bacillus* degraded DDT aerobically and faster than *E. coli* and *E. aerogenes*. Most of the metabolites in Table 1 were found in varying amounts after incubation for seven days, with all being readily identified after 30 days. To determine if the *Bacillus* followed the same pathway suggested by Wedemeyer for coliform bacteria, each metabolite in the table was indi-

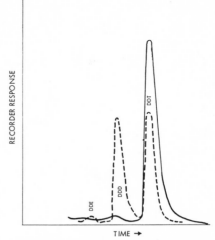

FIG. 2. Typical chromatograms obtained before and after 7 days of growth of *Escherichia coli* in trypticase soy broth plus 4.5% lactose containing 100 μg per milliliter of DDT. Symbols: before growth (———), after growth for 7 days (----).

vidually incubated with whole cells of *B. subtilis* in TSB. Extracts were analyzed after aerobic incubation for 7 and 30 days. Results indicated that DDT was degraded according to the same metabolic pathway suggested by Wedemeyer (15): DDT ⟶ DDD ⟶ DDMU ⟶ DDMS ⟶ DDNU ⟶ (DDOH) ⟶ DDA ⟶ DBP and DDT ⟶ DDE.

Amount of DDT degradation by *E. coli*, *E. aerogenes* and the *Bacillus* was affected by the kind of growth media. Since all results for these organisms were similar only those obtained for *E. coli* will be discussed.

Typical chromatograms before and after growth of *E. coli* in TSB, TSB plus 4.5% lactose (TSBL) and skimmilk containing 100 μg per milliliter of DDT are in Figures 1, 2 and 3. Most degradation of DDT occurred in TSB. After seven days of incubation most of the DDT had been converted to DDD and some DDE (Fig. 1). The other metabolites listed in Table 1 were not detected after 7 days but were found in varying amounts after 30 days of incubation.

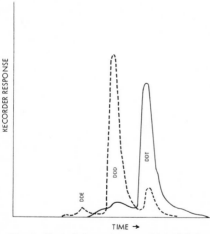

FIG. 1. Typical chromatograms obtained before and after 7 days of growth of *Escherichia coli* in trypticase soy broth containing 100 μg per milliliter of DDT. Symbols: before growth (———), after growth for 7 days (----).

In comparison to the aforementioned results, only about half of the DDT was degraded in TSBL (Fig. 2) after seven days of incubation. Additional time up to 30 days did not result

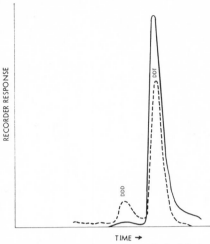

RECORDER RESPONSE

TIME →

Fig. 3. Typical chromatograms obtained before and after 7 days of growth of *Escherichia coli* in skimmilk or in trypticase soy broth plus 3% whole casein containing 100 μg per milliliter of DDT. Symbols: before growth (———), after growth of 7 days (----).

in a significant increase of DDD or DDE. None of the other metabolites in Table 1 was detected after 30 days of incubation. Reduction in amount of degradation probably was due to fermentation of lactose by bacteria which resulted in lowering pH of the medium. Wedemeyer (15) reported that the amount of metabolites formed was a function of the hydrogen ion concentration. Apparently as the pH is lowered, enzyme systems necessary for degradation are inhibited.

Unlike the two aforementioned broth media, very little degradation of DDT occurred in skimmilk or in TSB plus 3% whole casein (Fig. 3). Only small amounts of DDD could be detected after seven days of incubation and did not increase significantly after 30 days. Whole casein prevented heptachlor from inhibiting the growth of *S. aureus* (6). Apparently casein binds the pesticide and makes it unavailable for degradation by the bacteria. Addition of α-, β- and γ-caseins individually or a mixture of α- and β-caseins to TSB had some effect in reducing degradation; however, the reduction was not as great as that observed for whole casein.

Degradation of DDT in milk cannot be expected to occur, since casein appears to complex DDT and prevent degradation. High concentra-tions of acids resulting from utilization of sugars, also will prevent degradation of DDT. Degradation probably would occur in suitable liquid products without casein.

References

(1) Anonymous. 1967. Simplified thin-layer chromatography. Food Eng., 39: 76.

(2) Barker, P. S., and F. O. Morrison. 1964. Breakdown of DDT to DDD in mouse tissue. Canadian J. Zool., 42: 324.

(3) Barker, P. S., F. O. Morrison, and R. S. Whitaker. 1965. Conversion of DDT to DDD by *Proteus vulgaris*, a bacterium isolated from the intestinal flora of a mouse. Nature, 205: 621.

(4) Burchfield, H. P., and Donald E. Johnson. 1965. Guide to the Analysis of Pesticide Residues. Vol. I and II. Public Health Service, Office of Pesticides. Washington, D.C. 20201.

(5) Chacko, C. I., J. L. Lockwood, and M. Zabik. 1966. Chlorinated hydrocarbon pesticides; degradation by microbes. Science, 154: 893.

(6) Collins, J. A., and B. E. Langlois. 1968. Effect of DDT, dieldrin and heptachlor on the growth of selected bacteria. Appli. Microbiol., 16: 799.

(7) Hays, W. J. 1965. Review of the metabolism of chlorinated hydrocarbon insecticides, especially in mammals. Ann. Rev. Pharmacol., 5: 27.

(8) Hipp, N. J., M. L. Graves, J. H. Custer, and T. L. McMeekin. 1952. Separation of α-, β-, and γ-casein. J. Dairy Sci., 35: 272.

(9) Kallman, B. J., and A. K. Andrews. 1963. Reductive dechlorination of DDT to DDD by yeast. Science, 141: 1050.

(10) Langlois, B. E. 1967. Reductive dechlorination of DDT by *Escherichia coli*. J. Dairy Sci., 50: 1168.

(11) Langlois, B. E., A. R. Stemp, and B. J. Liska. 1964. Rapid cleanup of dairy products for analysis of chlorinated insecticide residues by electron capture gas chromatography. J. Agr. Food Chem., 12: 243.

(12) Matsumura, F., and G. M. Boush. 1967. Dieldrin: Degradation by soil microorganism. Science, 156: 959.

(13) Peterson, J. E., and W. H. Robison. 1964. Metabolic products of p,p'-DDT in the rat. Toxicol. Appl. Pharmacol., 6: 321.

(14) Pinto, J. D., M. N. Camien, and M. S. Dunn. 1965. Metabolic fate of p,p'-DDT (1, 1,1-trichloro-2,2-bis (p-chlorophenyl) ethane) in rats. J. Biol. Chem., 240: 2148.

(15) Wedemeyer, G. 1967. Dechlorination of 1,1,-1,trichloro-2,2-bis (p-chlorophenyl) ethane by *Aerobacter aerogenes*. Appl. Microbiol., 15: 569.

(16) Wedemeyer, G. 1968. Partial hydrolysis of dieldrin by *Aerobacter aerogenes*. Appl. Microbiol., 16: 661.

DDT-[14]C Metabolism by Rumen Bacteria and Protozoa in Vitro [1,2]

A. J. KUTCHES and **D. C. CHURCH**

Introduction

Numerous studies (8, 9) have shown that microbial systems possess the ability to convert DDT to DDD. Similarly, Braunberg and Beck (1) provided evidence that the major portion of DDD accumulation in rat feces was the result of microbial rather than mammalian metabolism. Recent work by Fries et al. (4), as well as earlier studies by Miskus et al. (6), have shown that whole rumen liquor is active in the conversion of DDT to DDD. While it is recognized that protozoa in symbiosis with bacteria contribute to the overall metabolic processes in the rumen, their metabolic role when separate from bacteria is reduced considerably (5) in volatile fatty acid production and dry matter digestion. The studies reported herein were initiated to determine the contribution of rumen protozoa, as well as whole rumen liquor and rumen bacteria, toward DDT metabolism in the rumen.

[1] Technical contribution 2939, Oregon Agricultural Experiment Station.

[2] Supported in part by PHS Grant ES 000239.

Experimental Procedure

Inoculum source. Rumen liquor was obtained 3 to 4 hr post-feeding from a fistulated steer on alfalfa-fescue hay, was strained through 4-layers of cheese cloth, transferred to a separatory funnel, and incubated for 1 to 1.5 hr at 39 C. During this period holotrich and larger entodiniomorph protozoa settled as a visible white layer at the bottom of the separatory funnel. This layer was withdrawn and resuspended in buffer[3] solution 3 to 4 times at 45 to 60 min intervals. Washing of the protozoal cells in this manner provided a suspension which was essentially free of plant debris.

After withdrawal of the protozoal fraction, the remainder of the debris-free rumen liquor was centrifuged at 121 $\times$ g for three minutes to remove the smaller protozoal cells consisting primarily of entodiniomorphs and remaining forage particles. These protozoal cells were then combined with the initial suspension.

Bacteria were harvested from the remaining supernatant by centrifugation at 41,300 $\times$ g for 20 min, resuspended in buffer, and the procedure repeated once more. Bacterial and protozoal suspensions were brought up to 300 ml volumes with buffer, and aliquots were then placed into the incubation vessels.

Portions of whole rumen liquor were saved for studies in which no special treatment was required.

With protozoal incubations, viable bacteria were controlled by the addition of 500 μg of buffered penicillin G and 500 μg of streptomycin sulfate per milliliter of buffer.

Incubation conditions. Incubations were in 100 ml tubes (29 $\times$ 200 mm) containing 15 ml of either whole rumen liquor or washed bacterial suspension or washed protozoal suspension and 1 μCi of DDT-[14]C (specific activity, 18.5 μg) and 40 μg per milliliter of carrier DDT. Incubations were under a continuous gas phase

---

[3] Composition of buffer solution: $NaHCO_3$, 4.90 g; $Na_2HPO_4 \cdot 7H_2O$, 3.50 g; KCl, 0.29 g; NaCl, 0.24 g; $CaCl_2$, 0.02 g; $MgSO_4 \cdot 7H_2O$, 0.06 g; Cysteine HCl, 0.25 g; resazurin (0.1%), 0.50 ml; clarified rumen liquor, 500 ml; distilled H_2O, to 1,000 ml.

of 5% CO_2–95% N_2 (v/v) which was freed of oxygen by passing the gas over hot copper filings. Incubations were maintained at 39 C for 0, 6, 18, 24 and 48 hr. Microbial activity was arrested in the zero hour vessels, which served as the controls, with 0.5 ml of 2 N H_2SO_4.

Extraction from incubation mixture. Following the incubation period, the vessel contents were centrifuged at 41,300 $\times$ *g* for 20 min and the supernatant fraction was separated from the cellular components by decanting. To each remaining fraction (cellular and supernatant) 25 ml of pentane and 5 ml of 0.1 N H_2SO_4 were added and periodically agitated for 24 hr. The flask contents were transferred to separatory funnels, and the lower layer was discarded. The remaining pentane layer containing DDT-^{14}C was withdrawn, evaporated to dryness in 50 ml Erlenmeyer flasks, and then brought to 1 ml volumes with *n*-hexane.

DDT and conversion product separation and quantitation. DDT-^{14}C and its conversion products were separated by thin-layer chromatography on prepared 20 $\times$ 20 cm Eastman Chromatogram sheets which were pretreated by immersing the entire sheet in methanol containing 1% $AgNO_3$ (7). Aliquots of 0.030 ml from the supernatant and cell fractions were spotted onto the pretreated chromatographic sheets and DDT-^{14}C along with its conversion products eluted with *n*-hexane in a standard chromatographic chamber.

Prior to separation of DDT and its conversion products, .020 to .030 ml of a prepared standard containing nonisotopic DDT, DDE and DDD (30 mg/ml each) were also spotted to assure positive identification since the conversion products of DDT stained weakly.

After migration in the solvent system, the chromatogram sheet was removed, dried, sprayed with 0.5% Rhodamine B and exposed to ultraviolet light to hasten development. Within 5 to 10 min the individual eluted zones became dark and were marked.

The individually marked areas were cut out and scraped into scintillation counting vials. To assure complete dispersion of carbon-14 in order to optimize the fluor system, 0.5 ml of methanol was added to dislodge the isotope from the silica gel. Fifteen milliliters of fluor solution were then added. The primary and secondary scintillators, respectively, in 1 liter of toluene were 5 g of 2,5-diphenyl-oxazole and 0.3 g of 1,4-*bis*-2-(5-phenyloxazolyl)-benzene. Samples were counted for 10 min or 10,000 counts per minute, whichever occurred first, by a Packard Tricarb (Model 3375) spectrometer. Disintegrations per minute (dpm) were calculated by automatic external standardization (AES) for quench correction.

Results and Discussion

Conversion of DDT-^{14}C to DDD by whole rumen liquor increased progressively with time, and at the end of the 48 hr incubation DDD accounted for 54% of the recovered activity (Table 1). Distribution of activity between DDT and its conversion products was comparable in both cell and supernatant fractions (Table 2) although over 95% of the recovered activity was associated with the cell fraction (cellular uptake).

Earlier work (4, 6) has shown a faster conversion of DDT to DDD than those in this study. It is likely that methods of incorporating the isotope into the reaction mixture together with gassing the reaction vessel have a bearing on the extent of DDT conversion.

Bacterial conversion of DDT-^{14}C to DDD increased with time at a faster rate than those shown to occur by whole rumen liquor, particularly at 6, 18 and 24 hr. However, at the end of the 48 hr incubation DDD was lower within the bacterial fraction and accounted for 48% of the recovered activity.

Cellular uptake of DDT-^{14}C (Table 1) was greater than 95% in all cases and suggests that, like biohydrogenation (3) reactions by rumen microbial fractions, conversion to DDD is an intracellular process. However, this does not preclude interaction of DDT with protein binding sites on the cell surface.

While the data are not conclusive, they suggest that DDT in the rumen is probably converted to DDD by reductive dechlorination rather than a two-step reaction with DDE serving as an intermediate product (8). The low distribution of activity associated with DDE at all times within the bacterial fraction would tend to support this view whereas results with whole rumen liquor are less clear.

34

TABLE 1. Conversion of DDT-^{14}C by cell fractions of rumen liquor.[a,b]

Source	Hour	Cellular uptake	Distribution of activity		
		(%)	(% total counts)		
			DDE	DDT	DDD
Whole rumen liquor	0[c]	96.76	2.57	89.19	8.23
	6	95.87	13.09	74.88	12.03
	18	99.17	16.82	64.99	18.17
	24	96.12	20.78	51.78	27.43
	48	98.98	2.54	43.86	53.60
Bacteria	0	93.80	1.54	86.03	12.42
	6	98.19	1.47	83.03	15.49
	18	98.96	2.75	63.53	33.71
	24	98.54	5.51	42.41	53.07
	48	97.44	4.53	47.61	47.85
Protozoa[d]	0	68.65	2.73	96.07	1.20
	6	97.54	1.85	93.65	4.49
	18	95.50	2.13	95.19	2.67
	24	94.79	4.44	88.19	7.37
	48	95.96	1.59	95.75	2.67

[a] All incubation vessels contained 15 ml and 1 μCi of DDT-^{14}C; vessels were incubated at 39 C for the times indicated.

[b] Means of two studies.

[c] Incubation contents were acidified with 1 ml 2 N H_2SO_4 at zero hour and incubated for 48 hr and served as the control for all times.

[d] Medium containing protozoal cells contained 500 μg per ml each of penicillin G and streptomycin sulfate.

35

TABLE 2. Supernatant distribution of DDT-^{14}C activity.[a]

Source	Hour	Distribution of activity		
		DDE	DDT	DDD
		(% total counts)		
Whole rumen liquor	6	3.73	76.86	19.40
	18	2.18	69.34	28.46
	24	11.90	56.55	21.42
	48	6.32	41.77	51.89
Bacteria	6	3.12	62.66	34.21
	18	2.43	31.35	66.21
	24	1.70	14.89	83.40
	48	7.89	23.68	68.42
Protozoa	6	4.23	90.47	5.29
	18	0.94	88.67	10.37
	24	19.67	76.22	4.09
	48	0.98	94.23	4.78

[a] Refer to footnotes in Table 1.

The distribution of DDT-^{14}C activity within protozoal suspensions consisting of both holotrichs and entodiniomorphs showed the inability of protozoa to convert DDT-^{14}C to either DDE or DDD. Even though cellular uptake of DDT-^{14}C was greater than 95% in all cases except at zero hour, the necessary mechanisms for DDT conversion appear to be lacking by mixed populations of rumen protozoa. While entodiniomorph protozoa participate in biohydrogenation reactions (3), their contribution toward DDT conversion appears limited as evidenced by results of this study with mixed suspensions of protozoa.

DDT-^{14}C activity of the controls (0 hr) for whole rumen liquor, bacteria, and protozoa were 89, 86, and 96% whereas activity of non-incubated DDT-^{14}C was greater than 96% as shown by thin-layer chromatography and radiochromatogram scan.

Recovery of carbon-14 from incubation vessels which contained whole rumen liquor ranged from 3 to 33%. Factors responsible for low recoveries are not clear although lipid material contained in whole rumen liquor was believed to contribute toward the low recoveries. Values for bacterial and protozoal carbon-14 recovery ranged from 43 to 72% in all except three cases in which recoveries were 14, 21, and 23%.

Important also in conversion of DDT to DDD is that porphyrins, under proper reducing conditions, possess the ability to convert DDT to DDD (6). In addition, it appears possible that reduced coenzymes and other metalloproteins could participate in these conversions. Likewise, porphyrins contained in rumen liquor (2) would undoubtedly contribute toward DDT conversion as well.

Results of this study, as well as others (4, 6), demonstrate that the rumen is an active site of DDT conversion. DDT conversion occurs by bovine liver perfusate (10) and, hence, makes available to the host two mechanisms whereby ingested DDT from contaminated ˙ feedstuffs may undergo detoxification.

References

(1) Braunberg, R. C., and V. Beck. 1968. Interaction of DDT and the gastrointestinal microflora of the rat. J. Agr. Food Chem., 16: 451.

(2) Caldwell, D. R., D. C. White, and M. P. Bryant. 1962. Specificity of the heme requirement for growth of Bacteroides ruminicola—a rumen saccharolytic bacterium. J. Dairy Sci., 45: 690.

(3) Chalupa, W. V., and A. J. Kutches. 1968. Biohydrogenation of linoleic-1-^{14}C acid by rumen protozoa. J. Animal Sci., 27: 1502.

(4) Fries, G. R., G. S. Marrow, and C. H. Gordon. 1969. Metabolism of o,p'- and p,p'-DDT by rumen microorganisms. J. Agr. Food Chem., 17: 860.

(5) Kutches, A. J. 1970. Influence of pesticides on rumen microbial metabolism. Ph.D. thesis, Oregon State University, Corvallis.

(6) Miskus, R. P., D. P. Blair, and J. E. Casida. 1965. Conversion of DDT to DDD by bovine rumen fluid, lake water and reduced porphyrins. J. Agr. Food Chem., 13: 481.

(7) Moats, W. A. 1966. Analysis of dairy products for chlorinated insecticide residues by thin-layer chromatography. J. Ass. Offici. Agr. Chemists, 49: 795.

(8) Plimmer, J. R., P. C. Kearney, and D. W. Von Endt. 1968. Mechanism of conversion of DDT to DDD by *Aerobacter aerogenes*. J. Agr. Food Chem., 16: 594.

(9) Wedemeyer, G. 1966. Dechlorination of DDT by *Aerobacter aerogenes*. Science, 152: 647.

(10) Whiting, F. M., S. B. Hagyard, W. H. Brown, and J. W. Stull. 1968. Detoxification of DDT by the perfused bovine liver. J. Dairy Sci., 51: 1612.

Dechlorination of DDT by Membranes Isolated from *Escherichia coli*[1]

ALLEN L. FRENCH and ROGER A. HOOPINGARNER

Kallman and Andrews (1963) were the first to demonstrate in vitro conversion of DDT to TDE (= DDD) by a microorganism (yeast). Following this report, interest grew in the role of microorganisms in the degradation of pesticides. Since DDT is extremely stable and has been extensively employed in the environment, many investigators have studied its metabolism.

Several species of bacteria isolated from animals, soil, and laboratory cultures convert DDT to TDE when cultured anaerobically (Baker and Morrison 1965, Stenersen 1965, Chacko et al. 1966, Wedemeyer 1966, Johnson and Goodman 1967, Plimmer et al. 1968). One case of DDT degradation by a fungus has been reported (Matsumura and Bousch 1968). Eighteen variants of soil-isolated *Trichoderma viride* were tested. Eight cultures produced both TDE and dicofol, 3 produced TDE, and 1 produced DDE (1,1-dichloro-2,2-bis (*p*-chlorophenyl) ethylene) and TDE

[1] Michigan Agricultural Experiment Station Journal Article no. 4741. This investigation was supported in part by Public Health Service Research Grant no. ES 00043 from the National Communicable Disease Center.

as their major metabolites.

Nonenzymatic degradation of DDT to TDE was reported by Castro (1964) and by Ecobichon and Saschenbrecker (1967). Castro demonstrated that dilute solutions of Fe^{++} can be oxidized at room temperature by alkyl halides, including DDT, to the corresponding Fe^{+++} halide complexes. Ecobichon and Saschenbrecker reported conversion of DDT to TDE, DDE, and unknown metabolites in frozen chicken blood that was repeatedly thawed.

Little is known about the mechanisms involved in the conversion of DDT to TDE by microorganisms. Wedemeyer (1966), employing sonerated *Aerobacter aerogenes* and selected metabolic inhibitors, proposed cytochrome oxidase as the enzyme responsible for this conversion. Plimmer et al. (1968), employing deuterated DDT, demonstrated that DDE was not an intermediate.

In this investigation we have utilized cellular components of the bacterium *Escherichia coli* to elucidate the metabolic processes involved in reductive dechlorination of DDT to TDE.

MATERIALS AND METHODS.—Cytoplasmic and particulate components of *E. coli* were separated by the method of Nagata et al. (1966). Cells were incubated with lysozyme '(Sigma Chemical Corp.) in buffered sucrose medium. The resulting protoplasts were harvested by centrifugation and subjected to osmotic shock. The particulate components were separated from the cytoplasm by centrifugation at (20,000 × g) for 20 min.

Metabolism studies were carried out in Warburg flasks containing 0.1 g of 15-μ glass beads coated with 350,000 dpm of DDT-^{14}C (uniformly ring labeled). The incubation components were maintained at 37°C under nitrogen. After being incubated 4 hr, the particulate and supernatant fractions were separated by centrifugation 20,000 × g for 20 min. DDT-^{14}C and its metabolites were extracted from membrane fractions with acetone homogenization, from aqueous fractions with hexane in separatory funnels. The extracts were combined and assayed for DDT and its metabolites.

DDT-^{14}C- and ^{14}C-labeled metabolites were identified by thin-layer and gas-liquid chromatography. Authentic unlabeled samples of DDT, o,p'-DDT, TDE, and DDE were spotted with 15,000 dpm of each extract on precoated silica gel H thin-layer chromatographic plates (Brinkmann Instrument Co.) and developed twice through 15 cm with n-hexane. Autoradiograms were obtained by exposing Kodak Medical X-ray film (no screen) to the plates for 4 days. The chromatograms were sprayed lightly with 0.1% alcoholic Rhodamine B and then covered with $NaCO_3$ solution onto filter paper to resolve the authentic standards. Quantification was accomplished

40

Table 1.—Effect of membrane and cytoplasm of *E. coli* on conversion of DDT.

Components	% [14]C found as DDT metabolites[a]				
	DDE	o,p'DDT	TDE	p,p'DDT	Un-known
Membrane only[b]	0.4	1.8	4.6	90.5	2.7
Membrane & cytoplasm[c]	.3	1.3	29.8	61.9	6.7
Cytoplasm only	.1	2.0	2.4	92.5	3.0
Boiled membrane & cytoplasm	.3	1.5	3.8	90.0	4.4
DDT-[14]C & buffer	.4	1.9	0.6	93.9	3.2

[a] Means of 2 experiments, 3 replicates each.
[b] The membrane fractions (3 ml aliquots) consisted of Tris buffer, at pH. 8.0, containing membranes at 25.0 mg/ml original dry weight of cells.
[c] The membrane fractions were resuspended in the cytoplasmic fraction at 25.0 mg/ml original dry weight of cells.

by collecting ^{14}C-labeled compounds from the effluent stream of the gas-chromatographic column. Samples were collected at 5-min intervals and assayed for ^{14}C-labeled effluent fractions and compared with those for authentic samples of TDE, o,p'-DDT, and DDE. A 2×4-mm glass column containing 80–100-mesh Gas Chrom-Q (The Anspec Co.) coated with 11.0% QF-1 and OV-17 (Applied Science Lab., Inc.) in a

Table 2.—Effect of exogenous Kreb's cycle intermediates and cofactors on DDT metabolism by membrane preparations of *E. coli.*

Components[a]	% [14]C found as DDT metabolites				
	DDE	DDT p,p'-	TDE	DDT	Un-known
Membrane only	0.4	1.8	4.6	90.5	2.7
Cofactors & intermediates[b]	0.2	1.3	2.2	93.2	3.1
Minus FAD	1.1	1.8	4.9	86.4	5.8
Minus ADP & PO$_4$	.0	1.2	5.9	89.1	3.8
Minus malate & pyruvate	.4	1.4	21.4	73.7	3.1
Minus NAD	.5	1.6	26.8	65.6	5.5
Minus NADP	.8	1.6	21.1	69.3	7.2
Control[c]	.1	0.4	1.3	94.7	3.5
Cofactors & intermediates (0.1×)	.1	1.6	7.9	86.3	4.1

[a] Each flask contained 3 ml of membrane fraction (25.0 mg/ml dry wt of cells).
[b] Two µmoles each of NAD, NADP, FAD, malate, pyruvate, and 0.1 µmole each of ADP and inorganic phosphate.
[c] Components consisted of 3 ml boiled membrane fraction plus exogenous cofactors. Malate and pyruvate were not added.

41

ratio of 1.3:1.0 was employed. Baseline separations of the authentic samples were achieved using a column temperature of 190°C, detector temperature of 200°C and nitrogen flow rate of 20 cc/min.

RESULTS.—Experiments were designed to ascertain the site of DDT metabolism in *E. coli*. Neither the cytoplasmic fraction alone nor the cytoplasmic fraction plus boiled membrane fraction displayed much ability to degrade DDT to TDE, but cytoplasmic fractions plus unboiled membranes produced substantial amounts of the TDE (Table 1). Thus the membranes of *E. coli* were the site of reductive dechlorination of DDT and the cytoplasm contained an essential factor (s).

In the presence of exogenous NAD (nicotinamide adenine dinucleotide), NADP (nicotinamide adenine dinucleotide phosphate), FAD (flavin adenine dinucleotide), ADP (adenosine diphosphate), inorganic phosphate, malate, and pyruvate the level of TDE recovered was not substantially different from that of the membranes alone (Table 2). When NAD, NADP or the Kreb's cycle intermediates (malate and pyruvate) were omitted from the incu-

Table 3.—Effect of NAD, FAD, ADP, and inorganic phosphate on DDT metabolism by membrane preparations of *E. coli*.

Components[a]	% ^{14}C found as DDT metabolites				
	DDE	*o,p'-* DDT	TDE	DDT	Un- known
NAD	0.3	1.3	7.7	87.4	3.3
FAD	.2	1.2	22.5	72.6	3.5
ADP & PO$_4$	.5	2.1	5.2	88.3	3.9
FAD, ADP & PO$_4$	.5	1.7	20.5	74.2	3.1
4X FAD, ADP & PO$_4$	.7	0.8	23.1	71.1	4.3
Cytoplasm, FAD, ADP & PO$_4$	.9	1.6	28.9	62.6	6.0
Cytoplasm & FAD	.1	1.3	26.2	68.8	3.6
Cytoplasm, ADP & PO$_4$	.5	1.3	27.9	65.1	5.2
FAD, ADP & PO$_4$ (aerobic)	.5	1.6	2.9	90.0	5.0

[a] Each flask contained 3 ml of membrane (25.0 mg/ml dry wt of cells).

bation mixtures, increases in TDE production occurred. Omission of ADP and inorganic phosphate or FAD from the incubation mixture did not result in increased TDE production. Significantly, substantial TDE production was achieved only in those membrane preparations containing FAD, ADP, and inorganic phosphate.

Since addition of exogenous ADP and inorganic phosphate or FAD did enhance TDE production by isolated membranes, experiments were conducted to determine the effects of these components singly or in combination (Table 3). Membrane fractions containing exogenous FAD, ADP, and inorganic phosphate or only FAD produced substantially more TDE than incubations containing membrane, ADP, and inorganic phosphate. Thus isolated membranes required the addition of exogenous FAD to obtain substantial TDE production.

The addition of FAD, ADP, and inorganic phosphate to membranes resuspended in the cytoplasmic fraction did not increase TDE production beyond that attained by membrane and cytoplasmic combinations only (Tables 1 and 3).

DISCUSSION.—After cellular lysis, neither the particulate membrane fraction nor the soluble fraction (cytoplasm) could produce significant amounts of TDE. However, if these fractions were combined, conversion of DDT to TDE occurred. After boiling the membrane fraction for 5 min and combining it with the cytoplasmic fraction, significant conversion did not occur. These results lend support to the argument that TDE production by microorganisms is enzymatically mediated and is not merely passive as reported by Castro (1964) and by Ecobichon and Saschenbrecker (1967). These observations and the fact that the addition of FAD to membrane fractions enhanced TDE production suggest that the capacity to metabolize DDT to TDE residues in the membranous portion of the bacterial cell and is not cytoplasmic. Since the cell walls were depolymerized and rendered soluble by the action of lysozyme, it most probably plays no direct role in reductive dechlorination of DDT.

The membrane fraction plus NAD, NADP, pyruvate, malate, FAD, ADP, and inorganic phosphate converted little DDT to TDE. However, if NAD, NADP or the 2 Kreb's cycle intermediates were deleted from the incubation components, significant increases in TDE production occurred. Thus the presence of potential Kreb's cycle activity inhibits TDE production under the conditions of this study.

The addition of FAD, ADP, and inorganic phosphate to membrane fractions incubated aerobically did not enhance TDE production. Thus FAD enhancement of TDE production is dependent on anaerobic conditions. This fact suggests that normally operating oxidative pathways preclude the reductive dechlorination of DDT.

43

Addition of malate, pyruvate, and cofactors of the Kreb's cycle to isolated membrane fractions did not enhance TDE production beyond that obtained by membranes alone. Since membrane preparations of *E. coli* are capable of metabolizing Kreb's intermediates (Mizuno et al. 1961, Gray et al. 1966, Cox et al. 1968) and contain the cytochromes b, a, a_2 and c (Gray et al. 1966), one would not expect the results obtained in this investigation. If reduced cytochrome a_3 (cytochrome oxidase) is indeed the enzyme responsible for the conversion of DDT to TDE (Wedemeyer 1966), deletion of major components of the Kreb's cycle should not enhance TDE production. On the contrary, their metabolism should contribute electrons to the cytochrome system and result in increased TDE production.

Four-fold increments of exogenous FAD added to membrane fractions failed to significantly increase TDE production beyond that obtained by the addition of 2 μmole amounts. This fact suggests that another factor (or factors) is limiting the rate of TDE production. Cytoplasmic stimulation of DDT reduction by membrane preparations was not increased by addition of exogenous FAD. The stimulating factor (or factors) that was present in the cytoplasmic fraction is unknown. The isolation and characterization of this factor (s) required for DDT reduction would contribute valuable information concerning the metabolic processes involved in DDT reduction.

Based on the results of this investigation the following posibilities are suggested. Reductive dechlorination of DDT occurs in the membranous portion of the bacterial cell and is not cytoplasmic in origin. It is stimulated by a component (s) in the cytoplasm and does not utilize electrons produced by the oxidation of Kreb's cycle intermediates and passed through the cytochrome system. Reductive dechlorination of DDT is dependent upon the reduction of FAD and occurs only under anaerobic conditions. The oxidation of endogenous substrates can produce the half-reduced form of FAD (FADH, a semiquinone) and may be the active moiety involved in the enzymatic reduction of DDT.

REFERENCES CITED

Baker, P. S., and F. O. Morrison. 1965. Conversion of DDT to DDD by *Proteus vulgaris,* a bacterium isolated from the intestinal flora of a mouse. Nature 205: 621–2.

Castro, C. E. 1964. The rapid oxidation of iron (II) porphyrins by alkyl halides. A possible mode of intoxication of organisms by alkyl halides. J. Amer. Chem. Soc. 86: 2310–1.

Chacko, C., I., J. L. Lockwood, and M. Zabik. 1966. Chlorinated hydrocarbon pesticides: degradation by microbes. Science 154: 893–5.

Cox, G. B., A. M. Snoswell, and F. Gibson. 1968. The use of a ubiquinone-deficient mutant in the study of malate oxidation in *E. coli.* Biochim. Biophys.

Acta 153: 1–12.

Ecobichon, D. J., and P. W. Saschenbrecker. 1967. Dechlorination of DDT in frozen blood. Science 156: 663–5.

Gray, C. T., J. W. T. Wimpenny, D. E. Huges, and N. Nossman. 1966. Regulation of metabolism in facultative bacteria. Biochim. Biophys. Acta 117: 22–32.

Johnson, B. T., and R. N. Goodman. 1967. Conversion of DDT to DDD by pathogenic and saprophytic bacteria associated with plants. Science 157: 560–1

Kallman, B. J., and A. K. Andrews. 1963. Reductive dechlorination of DDT to DDD by yeast. Science 141: 1050–1.

Matsumura, F., and G. M. Bousch. 1968. Degradation of insecticides by a soil fungus, *Trichoderma viride*. J. Econ. Entomol. 61: 610–2.

Mizuno, S., E. Yoshida, H. Takahashi, and B. Maruo. 1961. Experimental proof of a compartment of energy-rich-P" in a subcellular system from *Pseudomonas fluorescens*. Biochim. Biophys. Acta 49: 361–81.

Nagata, Y., S. Mizuno, and B. Naruo. 1966. Preparation and properties of active membrane systems from various species of bacteria. J. Biochem. 59: 404–10.

Plimmer, J. R., P. C. Kearney, and D. W. Von Endt. 1968. Mechanism of conversion of DDT to DDD by *Aerobacter aerogenes*. J. Agr. Food Chem. 16: 594–7.

Stenersen, V. H. J. 1965. DDT metabolism in resistant and susceptible stableflies and in bacteria. Nature 207: 660–1.

Wedemeyer, G. 1966. Dechlorination of DDT by *Aerobacter aerogenes*. Science 152: 647.

45

Chemical Analysis of DDT and Derivatives

Residues of Aldrin, Dieldrin, Chlordane, and DDT in Soil and Sugarbeets[1,2]

Jerome A. Onsager, H. W. Rusk, and L. I. Butler

Most of the irrigated land in the Pacific Northwest has been treated with persistent organochlorine insecticides, especially aldrin and DDT, for control of wireworms (Elateridae). Therefore, in 1965, at the Potato, Pea, and Sugarbeet Insects Investigations laboratory at Yakima, Wash., we wanted to determine the amount of residues of these insecticides that are currently present in raw sugarbeets or in dehydrated beet pulp, an important byproduct of processing that may constitute a significant fraction of the ration of domestic livestock. The rate of decay of insecticidal residues in soil has been reported by several authors, and the fate of residues during the processing of raw sugarbeets was studied by Walker et al. (1965). Also, the results of feeding known amounts of insecticide to livestock have been reported by several authors. However, we lacked definitive information about the relationship between the concentration of residues in soil and the concentration of residues in the sugarbeets grown in that soil; the present study was designed to provide that information.

METHODS AND MATERIALS.—*Procurement of Samples.* —On Mar. 31–Apr. 2, 1965, 6 rates each of aldrin, chlordane, and DDT were applied and disked 5–6 in. into unreplicated 40×50-ft plots of Umapine loam soil. All insecticides were applied as EC formulations diluted with water to a volume of 20 gal/acre. The

[1] In cooperation with the College of Agriculture, Research Center, Washington State University, Pullman, and the Utah-Idaho Sugar Co., Toppenish, Wash.

[2] Mention of a pesticide or a proprietary product does not imply endorsement of by the USDA.

rates of actual insecticide applied/acre were 10, 7.5, 5, 2.5, 1.25, and 0.625 lb of aldrin and 20, 15, 10, 5, 2.5, and 1.25 lb of chlordane or DDT.

Sugarbeets were then grown in these plots during 3 consecutive years, 1965–67. Each year the beets were seeded during early April and were harvested about 6 months later during early October. The soil was sampled at the time of each planting and each harvest by collecting 20 cores of soil, each 1 in. diam × 6 in. deep, from near the center of each treated plot and from the untreated check plots, 1 for each of the 3 insecticides. In addition, the soil in all plots that had been treated with aldrin was likewise sampled during late March 1968. The sugarbeets were sampled at time of harvest by selecting 10–12 beets at random from the central area of each treated or untreated plot.

Extraction of Samples.—The samples of soil were frozen, stored, thawed, screened, and mixed; then 100-g subsamples were sealed in bottles containing 200 ml of hexane and 100 ml of isopropanol, and the bottles were tumbled for 1 hr in a machine. Next, each mixture was filtered, the filtrate was extracted 3 times with distilled water, and the hexane fraction was then filtered through anhydrous sodium sulfate and refrigerated.

The crowns of the sampled beets were removed, and the roots were stored at 5°C. Later, the roots were brushed under running water to remove all adhering soil and quartered longitudinally with a band saw; 1/4 of each was retained. These samples were ground, and a 100-g subsample from each plot was placed in a plastic bag and quick-frozen. These subsamples were later blended with 200 ml of acetonitrile for 5 min at high speed in a Waring Blendor®. Anhydrous sodium sulfate was added, the mixtures were filtered, and 100-ml portions were transferred to 500-ml separatory funnels. To each funnel was added 200 ml of distilled water, 50 ml of saturated sodium sulfate solution, and 50 ml of hexane. After extraction, the hexane solution was removed and filtered through anhydrous sodium sulfate. Each separatory funnel and filter funnel was rinsed with 30 ml of hexane. The filtrates were evaporated just to dryness in a warm water bath with a stream of air, and the resulting residues were dissolved in measured volumes of hexane and refrigerated.

Cleanup of Samples.—The beet samples from 1965 did not require cleanup. Hexane solutions of beets from 1966 and 1967 were transferred to separatory funnels, and each container was rinsed 3 times with 10 ml of hexane. After the solutions were extracted successively with 15 ml of a 1:1 mixture of fuming

49

sulfuric acid and concentrated sulfuric acid, 15 ml of concentrated sulfuric acid, and 50 ml of water, they were dried by filtration through sodium sulfate on cotton. The filters and separatory funnels were each rinsed twice with 15 ml of hexane, and the solutions were evaporated to dryness and then dissolved in hexane, as previously described, and refrigerated.

Aliquots of the hexane extracts of beets (equivalent to 20 g of beets) from the aldrin and chlordane plots of 1966 and 1967 were shaken for 5 min with 1 g of a 1:1 mixture of Nuchar C-190 N® and Magnesol® and filtered through sodium sulfate on cotton. The flasks and funnels were each rinsed 4 times with 20 ml of a mixture of 10% distilled ethyl ether, 10% distilled benzene, and 80% distilled hexane. The filtrates were then evaporated to dryness

Table 1.—Average recoveries of aldrin, dieldrin, chlordane, and isomers of DDT from known quantities added to soil and sugarbeet roots. 1965–68.

| | Average% recovery[a] at concentrations of 0.025–5 ppm | | | | | |
	Aldrin	Dieldrin	Chlor-dane	DDE	o,p'-DDT	p,p'-DDT
Soil	87.6	102.2	97.2	101.4	102.8	115.4
Roots	81.4	86.2	86.5	89.9	65.3	76.8

[a] Corrected for the appropriate control samples.

and dissolved in hexane, as previously described, and refrigerated.

Analysis of Samples.—All residues were determined by electron-capture gas chromatography. During the 3 years of the project, several different columns commonly used for chlorinated insecticides were used in the gas chromatograph.

Known amounts of each insecticide were added to untreated control samples to determine the average percentage of recovery (Table 1) for each insecticide. All analytical data were corrected for average percentage of recovery and for the levels of "apparent" insecticide found in samples from the untreated control plots. In pretreatment soil samples from each plot, no chlordane or TDE was detected. However, pretreatment levels of DDE, o,p'-DDT, p,p'-DDT, aldrin, and dieldrin were 0.03, 0.03, 0.2, 0.004, and 0.03 ppm, respectively.

The qualitative validity of about 10% of the analyses (selected at random) was confirmed by determining the extraction p-values (Bowman and Beroza 1965).

Statistical Techniques.—The concentration of insecticide found in each soil sample was converted to a percentage of the concentration found in the same plot during April 1965, shortly after the treatments had been applied. A degradation curve was then calculated for each rate of each insecticide by using the techniques of regression and correlation analysis. With 1 exception, the calculated degradation curve for each of the 6 rates of each insecticide was significant at the 5% level of error. Analyses of covariance showed that the degradation curves for the 6 different concentrations of the same insecticide did not differ significantly in either slope or magnitude. This was rather conclusive evidence that each of the 6 curves was a sample of a common degradation curve that was independent of concentration. Therefore, all data for each insecticide could be pooled to obtain a precise estimate of the rate of degradation. For chlordane and DDT, the curves were calculated from 72 observations (2 subsamples × 6 concentrations × 6 dates of sampling). For aldrin and its metabolite dieldrin, the curves were calculated from 48 observations each.

The correlation between the concentration of residues in the soil at planting time and the concentration of residues in mature sugarbeets was then calculated for each insecticide. The curve was calculated from 36 observations (2 subsamples × 6 concentrations × 3 dates of sampling) for chlordane and DDT and from 12 and 24 observations for aldrin and dieldrin, respectively.

RESULTS AND DISCUSSION.—Table 2 gives the average residues of aldrin and dieldrin, DDT and metabolites, and chlordane in soils and sugarbeet roots.

Degradation curves appear in Fig. 1. The correlation coefficients for aldrin, aldrin + dieldrin, chlordane, DDT, and dieldrin alone were −0.92, −0.84, −0.91, −0.78, and −0.94, respectively. The half-life of aldrin was 3.1 months for 20 months, after which only about 1% of the initial concentration remained as aldrin. However, much of the aldrin was epoxidized to dieldrin, which had a much longer half-life. Consequently, the half-life of the residue of aldrin + dieldrin was 8.5 months for about the 1st 20 months (while significant quantities of aldrin were still present) but thereafter, the half-life of dieldrin alone was 29.7 months. The half-lives for chlordane and DDT were 14.3 and 22.9 months, respectively. All of these half-lives agreed closely with published data and were well within the ranges that have been reported by several investigators as summarized by Marth (1965) and Edwards (1966).

Residues of all insecticides in mature sugarbeets were directly proportional to the residues in the soil at the time of planting (Fig. 2). Sugarbeets apparently have a much greater affinity for aldrin than for dieldrin. In 1965 (the year aldrin was applied), most of the residue in beets must have been absorbed as aldrin and metabolized to dieldrin, because the con-

51

Table 2.—Average residues of aldrin, DDT, and chlordane in soil and sugarbeet roots. 1965-68.

Dosage (lb AI/acre)	Average residue (ppm) [a]									
	Soil sampled on —							Roots sampled on —		
	4/10/65	10/5/65	4/15/66	10/25/66	4/3/67	10/17/67	4/10/68	10/5/65	10/25/66	10/17/67
	Aldrin (aldrin and dieldrin combined)									
0.625	0.05	0.01	<0.01	0.01	<0.01	<0.01	<0.01	<0.01	<0.01	<0.01
1.25	.21	.06	.03	.03	.03	.02	.02	.04	<.01	<.01
2.5	.38	.11	.07	.05	.07	.04	.05	.06	.01	.01
5.0	.73	.39	.24	.11	.16	.09	.11	.13	.02	.01
7.5	1.84	.66	.82	.21	.27	.14	.24	.37	.03	.03
10.0	2.84	.97	.77	3.3	.52	.20	.29	.96	.05	.04
	DDT (DDE, *o,p'*—DDT and *p,p'*—DDT combined)									
1.25	1.18	.45	.19	.05	.16	.06		.02	<.01	.01
2.5	1.84	.76	.33	.07	.28	.20		.04	.01	.01
5.0	5.12	1.10	.74	.41	.64	.53		.08	.04	.04
10.0	9.94	2.36	1.13	1.12	1.57	1.31		.20	.10	.06
15.0	14.38	4.42	2.55	2.01	2.10	1.72		.35	.10	.11
20.0	14.47	5.32	3.32	3.77	3.18	2.00		.33	.17	.17
	Chlordane									
1.25	.25	.18	.22	.03	.17	<.01		.02	.01	<.01
2.5	.50	.67	.44	.14	.31	<.01		.08	.01	.02
5.0	1.08	1.28	1.05	.38	.72	.15		.37	.02	.05
10.0	2.90	2.90	2.45	.78	1.34	.33		.61	.10	.12
15.0	2.46	4.42	2.59	1.10	1.65	1.18		.73	.10	.17
20.0	6.62	4.14	3.88	2.25	2.74	2.25		1.12	.24	.26

[a] Corrected for control values and for recovery.

52

centration of dieldrin in mature beets averaged 614%
of the concentration of dieldrin in the soil at time
of planting. However, combined residues of aldrin
and dieldrin in beets averaged only about 18% of the
combined concentration in soil at planting (Fig. 2).
Conversely, in March of 1966 and 1967, 87% and
>99%, respectively, of the total combined residue in
soil was dieldrin. During those years, the combined

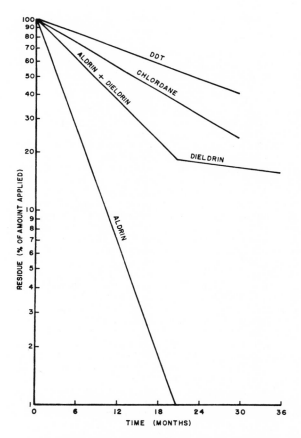

FIG. 1.—Rate of decay of aldrin, chlordan, DDT, and
dieldrin in Umapine loam.

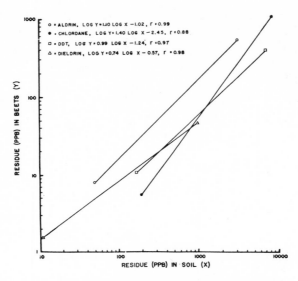

FIG. 2.—Residues of aldrin, chlordane, DDT, and dieldrin in mature sugarbeets and in soil at the time of planting (1965–67).

residue in beets was almost 100% dieldrin and averaged only 8.4% of the combined concentration in soil at planting (Fig. 2).

Residues of chlordane in mature sugarbeets averaged 9.6% of the concentration in the soil at the time of planting (Fig. 2). These data were more variable than data for the other insecticides. The variability is probably the result of analytical error and rounding of numbers, because chlordane consists of numerous isomers that produce different peaks on the gas chromatogram. The residues of chlordane in soil and beets included about 1.5% heptachlor, which is one of the normal constituents of technical chlordane (Ingle 1965). However, no heptachlor epoxide was detected in either soil or beets.

Within 6 months after application, the residue of DDT and its metabolites in soil had stabilized at 76% p,p'-DDT, 16% o,p'-DDT, and 11% DDE, and residues in sugarbeets averaged 73% p,p'-DDT, 17% o,p-DDT, and 11% DDE. Thus, beets apparently have an equal affinity for all 3 isomers. The total combined residue of DDT and metabolites in mature beets averaged 5.5% of the total concentration in the soil at the time of planting (Fig. 2).

These data suggest that small but measurable quantities of residue can now be detected in sugarbeets grown in soil treated with aldrin, chlordane, or DDT during the 1950's and early 1960's. However, the magnitude of residues currently resulting from normal past usage or present registered usage of the insecticides is expected to be very low and of questionable significance. For example, if sugarbeets are planted 2 years after a treatment with 5 lb of aldrin/acre, the residue in soil should be about 0.9 lb of dieldrin/acre or about 0.45 ppm. Residue in sugarbeets should then average about 0.024 ppm and the residue in dehydrated pulp from those beets will average about 10.6× that concentration (Walker et al. 1965) or about 0.25 ppm. A residue of 0.25 ppm in 4.5–8 lb of beet pulp fed daily to a 1200-lb animal amounts to ingestion of 0.9–1.6 ppb of insecticide.

REFERENCES CITED

Bowman, M. C., and M. Beroza. 1965. Extraction p-values of pesticides and related compounds in six binary solvent systems. J. Ass. Offic. Agr. Chem. 48: 943–52

Edwards, C. A. 1966. Insecticide residues in soils. Residue Rev. 13: 83–132.

Ingle, L. 1965. A monograph on chlordane. Velsicol Chemical Corp., Chicago. 88 p.

Marth, E. H. 1965. Residues and some effects of chlorinated hydrocarbon insecticides in biological material. Residue Rev. 9: 1–85.

Walker, K. C., J. C. Maitlen, J. A. Onsager, D. M. Powell, L. I. Butler, A. E. Goodban, and R. M. McCready. 1965. The fate of aldrin, dieldrin, and endrin residues during the processing of raw sugarbeets. USDA ARS 33–107. 8 p.

Systematic Studies on the Breakdown of *p,p'*-DDT in Tobacco Smokes

II. Isolation and Identification of Degradation Products from the   Pyrolysis of *p,p'*-DDT in a Nitrogen Atmosphere

N. M. Chopra and Neil B. Osborne

IN THE FALL of 1967 we embarked upon a systematic investigation into the breakdown of *p,p'*-DDT in cigarette main-stream and side-stream amokes.   This is the first study of its kind and is divided into three phases—the study of the breakdown of *p,p'*-DDT in a nitrogen atmosphere, in *p,p'*-DDT treated tobacco smokes, and in the cigarette main-stream and side-stream smokes.   So far we have written two papers on the results and mechanisms of the breakdown of *p,p'*-DDT in a nitrogen atmosphere, and in *p,p'*-DDT treated tobacco smokes (*1, 2*).

(1) N. M. Chopra, J. J. Domanski, and N. B. Osborne, *Beitr. Tabakforsch.*, **5**, 167 (1970).

(2) N. M. Chopra, *Proc. Second Intern. Congr. Pesticide Chem.*, Tel Aviv, Israel, 1971, in press.

In the first phase our objective was to study, from the break-down of *p,p'*-DDT in an inert atmosphere such as nitrogen, the mechanisms of the breakdown of *p,p'*-DDT at 900 °C— the temperature of the on-puff burning zone of cigarette (3). In this study the temperature was the only parameter which corresponded to the smoking condition of a cigarette. Our choice of an inert atmosphere turned out to be fortunate since we later on found (2) that the atmosphere where the pyrolysis of pesticide takes place is inert.

The results of our work on the first phase are described in our first paper (1) in Section 1. In the present paper we are re-porting the methods employed in the pyrolysis of *p,p'*-DDT, and the isolation and identification of its pyrolysis products. The products reported are: *p,p'*-DDT [2,2-di-(*p*-chloro-phenyl)-1,1,1-trichloroethane] and compounds with smaller molecular weight such as; *p,p'*-DDE [2,2-di-(*p*-chlorophenyl)-1,1-dichloroethylene], *p,p'*-TDE [2,2-di-(*p*-chlorophenyl)-1,1-dichloroethane], *p,p'*-DDM or *p,p'*-TDEE [2,2-di-(*p*-chloro-phenyl)-1-chloroethylene], *cis*- and *trans*-4,4'-dichlorostil-benes, *bis*-(*p*-chlorophenyl)methane, *bis*-(*p*-chlorophenyl)-chloromethane, α,p-dichlorotoluene (*p*-chlorobenzyl chloride), hexachloroethane, tetrachloroethylene, trichloroethylene, car-bon tetrachloride, chloroform, and dichloromethane.

EXPERIMENTAL

Materials. All solvents used were of "pure" grade, and were distilled before use. ("Pure" and "Puriss" grades refer to the quality of the reagent as mentioned on the rea-gent bottle. "Puriss" grade was 99.9%+ pure.) *p,p'*-DDT used for pyrolysis was of "Puriss" grade, and was purchased from Aldrich Chemical Co.

ALUMINA AND FLORISIL. Alcoa chromatographic alumina F-20 was purchased from Aluminum Company of America, and Florisil (mesh, 100–200) was purchased from Fisher. These were activated as described in the text.

REFERENCE COMPOUNDS. *p,p'*-DDT (99.9% + pure) and *p,p'*-DDE (99.8% pure) were obtained from Geigy Chemical Corp., and *p,p'*-TDE (pure) was obtained from Rohm and Haas.

4,4'-Dichlorobiphenyl was purchased from Chemical Procurement Laboratories. It was crystallized from pentane. The purified product had an IR spectrum identical with that of 4,4'-dichlorobiphenyl as reported in literature (4).

Bis-(*p*-chlorophenyl)methane, α,p-dichlorotoluene, and hex-

---

(3) G. P. Touey and R. C. Mumpower, *Tobacco Sci.*, **1**, 33 (1957).

(4) "Sadtler Standard Spectra," Midget ed., Sadtler Res. Lab., Philadelphia, 1962, No. 15044.

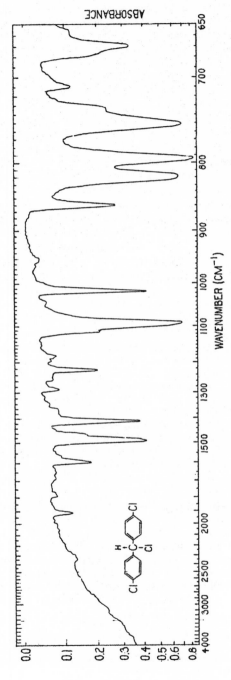

Figure 1. Infrared spectrum of *bis*-(*p*-chlorophenyl)chloromethane in KBr

58

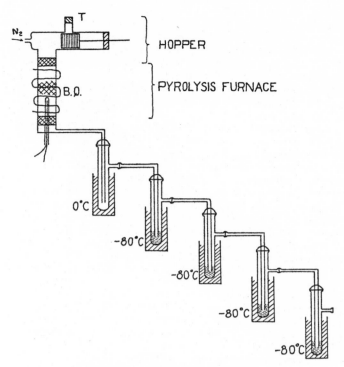

Figure 2. Pyrolysis apparatus

achloroethane were purchased from Eastman Organic Chemicals. They were crystallized from pentane. Their purified products had IR spectra identical with those of *bis-(p*-chlorophenyl)methane (*5*), *α,p*-dichlorotoluene (*6*), and hexachloroethane (*7*), respectively.

Carbon tetrachloride, chlorobenzene, chloroform, dichloromethane, tetrachloroethylene, and trichloroethylene

(5) M. B. Abou-Donia and D. B. Menzel, *J. Assoc. Offic. Anal. Chem.*, **51**, 1247 (1968).
(6) "Sadtler Standard Spectra," Midget ed., Sadtler Res. Lab., Philadelphia, 1962, No. 626.
(7) *Ibid.*, No. 4546.

were of the purest grade available in the market. They were tested for purity, and purified wherever necessary.

p,p'-DDM (*p,p'*-TDEE) was synthesized from pure *p,p'*-TDE (20 g) by refluxing it with excess of alcoholic KOH for 24 hours [*cf.* Haller *et al.* (*8*)]. The reaction product was crystallized from isopropanol. Its melting point, 66–67 °C and IR spectrum were identical with those reported for *p,p'*-DDM (*5*).

Bis-(*p*-chlorophenyl)chloromethane was prepared by passing dry HCl through a solution of *p,p'*-dichlorobenzhydrol (5 g) in benzene (50 ml) for 6 hours. The benzene solution was then evaporated to dryness under reduced pressure. The resulting yellowish oily residue solidified on standing. It was dissolved in pentane and treated with activated charcoal. The solution was then filtered and concentrated when white crystals of the crude product appeared. These crystals were recrystallized from pentane to give pure bis-(*p*-chlorophenyl)chloromethane, mp 62–63 °C (*9*). *Anal.* Calculated for $C_{13}H_9Cl_3$: C, 57.4; H, 3.32; Cl, 39.22. Found: C, 57.27; H, 3.35; Cl, 39.10. Its IR spectrum is shown in Figure 1.

Trans-4,4'-dichlorostilbene was synthesized according to the method of Hoffmann and Rathkamp. The melting point, 176.5–177 °C, and the IR spectrum of the purified product was identical with that of trans-4,4'-dichlorostilbene (*10*).

Methods. PYROLYSIS OF *p,p'*-DDT. The pyrolysis apparatus is shown in Figure 2. It consisted of three sections: a hopper unit, the pyrolysis tube, and traps. The hopper unit and traps were made of glass, and the pyrolysis tube was made of quartz. All the connections in the apparatus were glass to glass.

p,p'-DDT (10 g) was introduced into tube (T) of the hopper and stoppered. From there it was introduced into the pyrolysis tube (PT) with a glass piston. From the other end of the hopper, nitrogen was introduced into the tube at the rate of 150 ml/min. The pyrolysis tube, 50 cm × 2.5 cm in diameter, had an indentation in the middle of which a 2-cm thick bed of broken quartz pieces (BQ) rested. The pyrolysis tube was heated in a muffle furnace to 900 °C. Small quantities of *p,p'*-DDT were dropped on a red hot broken quartz bed over a period of about 2 hours. The escaping pyrolysis products were trapped in five traps. The first trap was empty and was cooled to 0 °C. The other traps contained about 30 ml of pentane, and were cooled to −80 °C.

(8) H. L. Haller, P. D. Bartlett, N. L. Drake, M. S. Newman, S. J. Cristol, C. M. Eaker, R. A. Hayes, G. W. Kilmer, B. Magerlein, G. P. Muller, A. Schneider, and W. J. Wheatley, *J. Amer. Chem. Soc.*, 67, 1591 (1945).
(9) J. F. Norris and C. Banta, *ibid.*, **50**, 1807 (1928).
(10) D. Hoffmann and G. Rathkamp, *Beitr. Tabakforsch.*, **4**, 201 (1968).

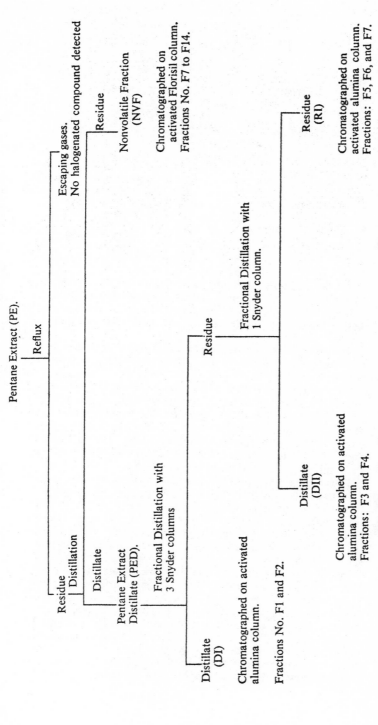

Figure 3. Procedure for isolation of pyrolysis products of p,p'-DDT

61

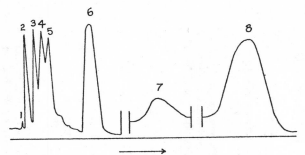

Figure 4. GLC chromatogram of PED on a 3% SE 30
column

1. Pentane
2. Dichloromethane
3. Chloroform
4. Carbon tetrachloride

5. Trichloroethylene
6. Tetrachloroethylene
7. Chlorobenzene
8. Hexachloroethane

After the pyrolysis operation was over, the pentane solutions
in the traps were combined and the residue left in each trap
was extracted with 20 ml of pentane. The pentane solutions
and extracts were combined and filtered, and the filterate
(PE) processed as shown in Figure 3.

PE was worked up into three fractions: Pentane Reflux
Fraction (PRF), Pentane Extract Distillate (PED), and Non-
volatile Fraction (NVF).

PE was refluxed with ice-cold water passing through the
reflux condenser. The escaping gases were trapped in a
pentane trap at −80 °C to give PRF. The PE, after the
removal of PRF, was distilled on a water bath at 50 °C to
give a viscous yellowish brown solid residue, NVF (0.155 g),
and the distillate PED. The gas chromatograms of the
PED and NVF are shown in Figures 4 and 5.

The chromatograms show that excluding the solvent there are
7 substances present in the PED and 20 substances in the NVF.

Isolation of Pyrolysis Products. PED was slowly
fractionally distilled in two stages. In the first stage, three
30-cm long Snyder fractionating columns were used and $^1/_3$
volume of the distillate (DI) was collected; and in the second
stage the remaining liquid in the flask was fractionally distilled three times with only one Snyder column till about 10
ml of the residue was left in the flask. The distillate was designated as DII, and the residue as RI. DI, DII, and RI were
gas chromatographically found to contain 2, 2, and 4 substances, respectively. DI was concentrated by fractional
distillation with three Snyder columns: the concentrate
consisted of the first 25 ml of the distillate. DII was concentrated by fractional distillation with four Snyder columns:
the concentrate, about 10 ml, was left in the flask.

Figure 5. GLC chromatogram of NVF on a 3% SE 30 column

1. Pentane
2. Hexachloroethane
3. α,p-Dichlorotoluene
4. p,p'-Dichlorobiphenyl
5. Bis-(p-chlorophenyl)methane
6. Cis-p,p'-dichlorostilbene
7. Bis-(p-chlorophenyl)chloromethane
8. p,p'-DDM
9. p,p'-DDE and trans-p,p'-dichloro-stilbene
10. p,p'-TDE
11. p,p'-DDT

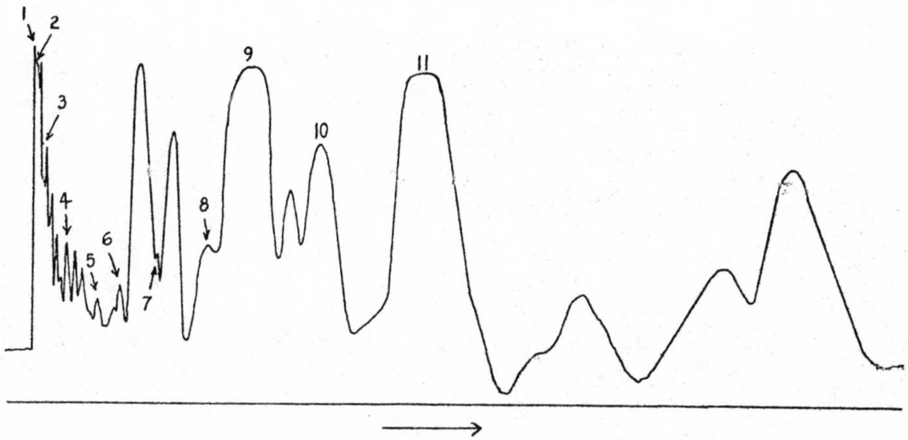

63

Concentrated DI and DII, RI, and NVF were chromatographed on suitable columns equipped with a fraction collector. In all the chromatographs, the rate of flow of the eluent was 1 ml per minute.

Concentrated DI (5-ml aliquot) was chromatographed on a 25 cm $\times$ 1.5 cm diameter activated alumina column (alumina was activated at 110 °C for 16 hours) with pentane as an eluent. Sixty 3-ml fractions were collected. These fractions, when examined gas chromatographically, showed the presence of two substances. The first substance appeared in fractions 13–17, and the second in fractions 38–45. These fractions were combined to give F1 (fractions 13–17), and F2 (fractions 38–45).

Concentrated DII (2-ml aliquot) was chromatographed on a 60 cm $\times$ 1.5 cm diameter activated alumina column (alumina activated at 250° for 16 hours) with pentane as an eluent. One hundred 3-ml fractions were collected. These fractions, when examined gas chromatographically, were found to contain two substances spread over fractions 25–41 and 45–65, respectively. These fractions were combined to give F3 (fractions 25–41), and F4 (fractions 45–65).

RI (2-ml aliquot) was chromatographed on a column identical with that used for DII. Pentane was used as an eluent and one hundred and fifty 2-ml fractions were collected. When examined gas chromatographically fractions 25–45 60–100, and 130–150 were found to contain one, three, and two substances, respectively. These fractions were combined to give F5 (fractions 20–45), FA (fractions 60–100), and FB (fractions 130–150). FA was carefully fractionally distilled with one Snyder column until 5 ml of liquid was left in the distillation flask. The distillate was examined gas chromatographically and was found to contain the same substance as found in F4. The residue was rechromatographed on a 60 cm $\times$ 1.5 cm diameter activated alumina column (alumina activated at 250 °C). Fractions 25–50 were found to contain the same substance as in F5. Fractions 75–100 were found to contain one substance. They were combined to give F7. Fractions 51–74 contained two substances one of which was present in traces. This substance was found to be the same as that present in F7. The combined fractions 51–74 were designated as F6. FB was similarly rechromatographed to give F6 and F7.

NVF (0.15 g) was chromatographed on a 50 cm $\times$ 5 cm diameter activated Florisil column (Florisil activated at 110 °C for 16 hours) with 10 liters of hexane followed by 8 liters of 1% ether in hexane as eluent. Eight hundred 20-ml fractions were collected. These fractions, when examined gas chromatographically, showed that a partial separation of the various constituents of the NVF was achieved. Fractions containing the same dominant single substance were combined and repeatedly rechromatographed till fractions containing the single dominant substance were obtained. Thus fractions 18–25, 35–50, 175–270, 310–335, 425–500, 600–625, 690–720, and 770–800 gave fractions F7, F8, F9,

64

F10, F11, F12, F13, and F14. The F7 obtained contained the same substance as the F7 from RI.

Methods of Detection. Gas Chromatography. A Micro Tek MT-220 gas chromatograph equipped with a ^{63}Ni electron capture detector, and a Dohrmann Model C-200 microcoulometer with a Model S-100 combustion unit were employed in this study. The three columns A, B, and C used were 6 ft $\times$ $^{1}/_{4}$-inch diameter glass columns. Column A was packed with 3% SE 30 on 80–90 mesh Chromoport XXX; column B with 5% SE 30 on 80–90 mesh Chromoport XXX; and column C with 20% Carbowax 20 M on 80–100 mesh Chromoport XXX. The column temperatures and gas flow varied depending upon the compound being investigated.

Colorimetric Tests. Colorimetric tests were done only on the constituents of the PED. Since all of the compounds were in pentane solutions, tests used in their detection had to be modified from what they were reported in literature. For chlorobenzene, a new test was developed in our laboratory. The colorimetric tests employed are as follows.

Test 1. Dichloromethane was detected by a method based on that by Gronsberg (*11*). A mixture of 1 ml of the sample and 2 ml of 20% ethanolic KOH was refluxed for 1 hour. The reaction mixture was cooled and then added to 3 ml of sulfuric acid containing 0.1–0.2 g of chromotropic acid. A violet color indicated the presence of dichloromethane (detection limit 5 μg). Carbon tetrachloride, chloroform, 1,2-dichloroethane, 1,1,2,2-tetrachloroethane, tetrachloroethylene, and trichloroethylene gave no color.

Test 2. Chloroform and trichloroethylene were tested by a modification of Fujiwara test as reported by Feigl (*12*). Five drops each of 20% NaOH and pyridine, and 0.5 ml of the test solution were heated at 100 °C for 1–2 minutes. To the red colored solution which was formed 3–4 crystals (*ca.* 1 mg) of benzidine hydrochloride were added and the solution acidified with acetic acid. Chloroform (detection limit, 2 μg) and trichloroethylene (detection limit, 5 μg) gave a violet color with this method, while dichloromethane (detection limit, 500 μg) and carbon tetrachloride (detection limit, 500 μg) gave a pink color. Tetrachloroethylene and hexachloroethane gave no color.

Test 3. Chloroform and trichloroethylene were also tested by a modification of Brumbaugh and Stallard's method (*13*). A mixture of 1 ml each of aniline and pyridine were

(11) E. Sh. Gronsberg, *Tr. Po. Khim i Khim Tekhnol*, **1**, 131 (1964); *Chem. Abstr.*, **62**, 996*h*, and **65**, 6303*e*.

(12) F. Feigl, "Spot Tests in Organic Analysis," 7th ed., Elsevier, New York, N. Y., 1966.

(13) J. H. Brumbaugh and D. E. Stallard, *J. Agr. Food Chem.*, **6**, 465 (1958).

refluxed with 2 ml of the sample at 135–140 °C for 15 minutes, and then 2 drops of 5% methanolic KOH were added. Chloroform and trichloroethylene gave a yellow-orange color. Tetrachloroethylene gave the color on further refluxing, while carbon tetrachloride and dichloromethane gave no color. The detection limits both for chloroform and trichloroethylene were 50 μg.

Chloroform was distinguished from trichloroethylene by heating the aniline–pyridine–sample mixture at 50 °C. Under these conditions only chloroform gave the color.

Test 4. Carbon tetrachloride and chloroform were detected by a modification of Belyakov's method (14). One milliliter each of pyridine and the test sample, and 2 drops of 1.0N NaOH were refluxed at 100 °C till a pink or red color developed (maximum time, 30 minutes). The colored solution was cooled and 3–4 drops of aniline and 1 ml of glacial acetic acid were added. Carbon tetrachloride and chloroform gave a yellowish orange color (detection limit for each, 5 μg). Dichloromethane gave no color.

Test 5. Chlorobenzene was detected by its conversion to aniline with potassium amide in liquid ammonia according to the method of Chopra and Domanski (15). Potassium metal (50 mg) was added to a dry 3-neck flask connected to nitrogen and ammonia tanks, and fitted with a dropping funnel and a Dewar condenser containing solid CO_2 and acetone. Ammonia was passed through the flask till about 50 ml of liquid ammonia was collected in the flask. A small amount of ferric nitrate was added as a catalyst. The reaction mixture was stirred vigorously till the original blue color of the reaction mixture turned gray. A pentane solution of the sample (ca. 10 ml) was then added. After 5 minutes the reaction was quenched by adding an excess of ammonium bromide, and liquid ammonia was allowed to evaporate. The residue, now containing aniline, was extracted with ether and the ether extract concentrated to 0.5 ml, and chromatographed on silica gel ITLC plates along with aniline as reference, and with chloroform as a mobile phase. The chromatograms were developed by spraying them with a solution containing 2 g α-naphthol, 10 ml of 25% triethylamine, and 90 ml of methanol, followed by the exposure of the chromatograms to nitric oxide fumes. The presence of chlorobenzene in the test sample was indicated by a spot of same color and R_f value as the reference spot.

(14) A. A. Belyakov, *Zav. Lab.*, **23**, 161 (1957); *Chem. Abstr.*, **51**, 17606c.
(15) N. M. Chopra and J. J. Domanski, North Carolina Agricultural and Technical State University, Greensboro, N. C. 27411, unpublished data, 1969.

Table I. Identification of Various Pyrolysis Products in PED from the Pyrolysis of p,p'-DDT in a Nitrogen Atmosphere

Fraction No.	No. of compounds present	Reference compound	Results of cochromatography with reference compounds and with E.C. and microcoulometric detectors			Results of colorimetric tests and their comparison with reference compounds					Identification of the unknown compound
			Column A	Column B	Column C	Test 1	Test 2	Test 3	Test 4	Test 5	
F1	One	Dichloromethane	No separation	No separation	No separation	violet SARC[a]	Pink SARC[a]	No color SARC[a]	No color SARC[a]	...	Dichloromethane
F2	One	Chloroform	No separation	No separation	No separation	No color SARC	violet SARC	yellow orange SARC	Yellow orange SARC	...	Chloroform
F3	One	Carbon tetrachloride	No separation	No separation	No separation	No color SARC	Pink SARC	No color SARC	yellow orange SARC	...	Carbon tetrachloride
F4	One	Trichloroethylene	No separation	No separation	No separation	No color SARC	violet SARC	yellow orange SARC	...	...	Trichloroethylene
F5	One	Tetrachloroethylene	No separation	No separation	No separation	No color SARC	No color	yellow orange SARC	...	...	Tetrachloroethylene
F6	Two	Chlorobenzene Hexachloroethane (see F7, Table 2)	No separation	No separation	No separation	...	...	...	...	Orange spots with R_f SARC	Chlorobenzene

[a] Same as reference compound.

67

Table II. Identification of Various Pyrolysis Products in NVF from the Pyrolysis of *p,p'*-DDT in a Nitrogen Atmosphere

Fraction No.	No. of compounds present	Reference compound	Results of cochromatography with reference compounds and with E.C. and microcoulometric detectors		Comparison of the IR spectra of the unknown with that of the reference	Identification of the unknown compound
			Column A	Column B		
F7	One	Hexachloroethane	SARC[a]	SARC[a]	SARC[a]	Hexachloroethane
F8	One	α,p-Dichlorotoluene	SARC[a]	SARC[a]	SARC[a]	α,p-dichlorotoluene
F9	One	*p,p'*-Dichloro-biphenyl	SARC[a]	SARC[a]	SARC[a]	*p,p'*-dichlorobiphenyl
F10	One	Bis-(*p*-chlorophenyl)-methane	SARC[a]	SARC[a]	SARC[a]	Bis-(*p*-chlorophenyl)methane
F11	One	Bis(*p*-chlorophenyl)-chloromethane	SARC[a]	SARC[a]	SARC[a]	Bis-(*p*-chlorophenyl)-chloromethane
F12	One	*p,p'*-DDE	SARC[a]	SARC[a]	SARC[a]	*p,p'*-DDE
F13	One	*p,p'*-TDE	SARC[a]	SARC[a]	SARC[a]	*p,p'*-TDE
F14	One	*p,p'*-DDT	SARC[a]	SARC[a]	SARC[a]	*p,p'*-DDT

[a] Same as reference compound.

RESULTS

The results of the experiments for the isolation and identification of various pyrolysis products obtained from the pyrolysis of p,p'-DDT in a nitrogen atmosphere are shown in Tables I and II. Table I shows the compounds present in the PED, and Table II, in the NVF.

The two tables confirm the presence of 14 compounds in the p,p'-DDT pyrolysis products. Three more compounds: p,p'-DDM, and *cis*- and *trans*-dichlorostilbenes were not isolated since they were present in very small amounts. However, their presence was shown by their cochromatography with reference compounds on columns A and B with E.C. and microcoulometric detectors.

ACKNOWLEDGMENTS

The authors thank Dr. B. R. Ray and Allan E. Foote for the work they did at the early stages of the research; Geigy Chemical Corporation and Rohm & Haas for the gift of pure p,p'-DDT and p,p'-TDE, respectively; Dr. H. P. Hermanson for the art work; and the Council for Tobacco Research–USA for the Research Grant which has made these investigations possible.

THE MECHANISM OF THE CHLORIDE ION-PROMOTED DEHYDROCHLORINATION

OF DDT AND RING-SUBSTITUTED ANALOGUES IN ACETONE

D. J. McLennan and R. J. Wong

Department of Chemistry, University of Auckland, Auckland, New Zealand.

Some controversy has arisen regarding the mechanism of bimolecular elimination reactions promoted by halide ions in dipolar aprotic solvents and by thiolate anions, both of which are weakly basic towards hydrogen (in the thermodynamic sense) but are strong carbon nucleophiles (in the kinetic sense). One view is that such eliminations proceed through the same type of transition state (I) as do normal E2 reactions promoted by conventionally strong bases,[1] while another is that partial covalent attachment of the nucleophile to C_α as well as to the β-hydrogen is important (II).[2]

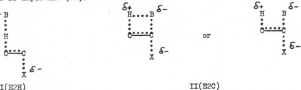

I(E2H)   II(E2C)

With one important exception,[1d] evidence that has hitherto been cited in favour of either the E2H or E2C mechanisms appears to be ambiguous to some extent, and is capable of being reasonably explained in terms of the alternative hypothesis. For instance, the fact that the S_N2 reactivities of a series of halide and thiolate ions towards cyclohexyl tosylate parallel their elimination reactivities towards this substrate has been interpreted in terms of the S_N2 and olefin-forming transition states being similar, with the latter necessarily having E2C character.[2c] However, the reactivities of the strong bases OEt^-, OMe^- and OH^- in their undoubted E2H reaction with 2-phenethyl bromide also parallel their S_N2 reactivities (but do not correlate with their hydrogen basicities).[1f]

The principal source of the ambiguities is that substrates which readily eliminate with halide ions in dipolar aprotic solvents are secondary and tertiary halides and arenesulfonates, which, when E2H conditions are employed (strong base), pass through a paenecarbonium transition state (III).[1a] Thus a similar pattern of results would be expected for either II or III as the

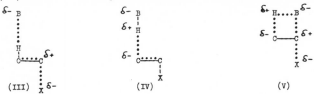

(III)   (IV)   (V)

transition state, namely Saytzeff orientation,[2a] little sensitivity to acidifying β-substit-uents,[2e] low Hammett ρ constant,[3] low deuterium isotope effect[4] and solvent effects exempli-fying "loose" transition states with much negative charge residing on B and X.[2g] A good case can be made for the transition states of halide-promoted eliminations of substrates so far studied being E2H and having somewhat more paenecarbonium character than the corresponding ones formed under undisputed E2H conditions. This would account for several notable differences in behaviour and is intelligible in view of the relative strengths of Hal....H and RO....H bonds.[1a]

DDT (VI; R = Cl) is dehydrochlorinated under E2H conditions (OEt$^-$/EtOH) clearly via a paene-carbanion transition state (IV).[5,6] Exploratory work has shown that dehydrochlorination occurs when n-Bu$_4$NCl in acetone containing 2,6-lutidine is used as the reagent, and that the kinetics are second-order.[7] Thus it is the only substrate known so far that undergoes E2H elimination through a paenecarbanion transition state and also undergoes halide-promoted elimination.

We have obtained the Hammett ρ constant for the reaction series:

$$(p\text{-RC}_6\text{H}_4)_2\text{CHCCl}_3 \xrightarrow{\text{n-Bu}_4\text{NCl}} (p\text{-RC}_6\text{H}_4)_2\text{C=CCl}_2 \;+\; \text{HCl}$$

$$\text{VI}$$

and kinetic results are shown in the Table. A plot of log k_2 vs. 2σ is linear (correlation

TABLE: The $(p\text{-RC}_6\text{H}_4)_2\text{CHCCl}_3$/n-Bu$_4$NCl Reaction[a] in Acetone Containing 2,4,6-Collidine[b] at 65°.[c]

R =	Me	H	Cl	Br	NO$_2$
$10^5 k_2$ (litre mole^{-1} sec.$^{-1}$)[d]	7.62	15.0	83.8	113	1430

[a]The requisite olefinic product has been isolated in good yield for R = Cl. [b][Coll.] has no effect on k_2. [c]Analysis by acid-base titration using a Radiometer automatic titrator. Potent-iometric titration for Cl$^-$ gave concordant results. [d][DDT]~0.02M; [n-Bu$_4$NCl]~0.01M; [Coll]~0.03M. Mean of two or more runs.

coefficient = 0.992; standard deviation = 0.13) and the ρ constant is 1.23. We contend that this value is too high to allow for the operation of an E2C mechanism, and argues in favour of an E2H transition state.

An E2C transition state resembles an S$_N$2 transition state except that in the former a little C$_\beta$— H bond loosening has occurred, although little β-carbanionic character is visualised.[2e] The ρ value for the S$_N$2 series ArCH$_2$CH$_2$Cl/I$^-$ in acetone[8] is 0.59, and those for the ArCH$_2$CH$_2$OTs/OEt$^-$ S$_N$2 series are 0.67 (EtOH) and 0.53 (t-BuOH).[9] One would expect ρ for E2C reactions of β-aryl-ethyl compounds to be somewhat higher than this, but hardly around twice as great.

Our ρ value is significantly smaller than that found for the DDT E2H dehydrochlorination using OEt$^-$/EtOH[5] (2.46 at 40.2°, extrapolating to approximately 2.27 at 65°) but this is to be expected in view of the difference in basic strengths of the hydrogen nucleophiles. The trans-ition state for the chloride-promoted dehydrochlorination of DDT thus appears to be E2H, and closer to "central" than is that for the OEt$^-$/EtOH reaction. C$_\beta$— H and C$_\alpha$— Cl bond-breaking will have made roughly equal progress[1a] but there must still be a significant negative charge density at C_β. In fact, ρ is larger than that observed for the clearly E2H reaction of $ArCH_2CMe_2Cl$ with $OMe^-/MeOH$.[10] S_N2-like interaction between the nucleophile and C_α of DDT will be energetically discouraged by the other halogens on C_α[11] and even the extremely strong carbon nucleophile, PhS^-, does not undergo S_N2 reaction with DDT.[6]

A second important piece of evidence in favour of the E2H mechanism is that $(p-ClC_6H_4)_2CHCHCl_2$ (DDD) is not visibly dehydrochlorinated using $n-Bu_4NCl$ in acetone under conditions where DDT is at least 50% decomposed. This is intelligible in terms of an E2H process, in which α-halogens accelerate,[1e,5] but not in terms of an E2C mechanism involving important S_N2-like interactions, since α-halogens retard S_N2 reactions.[11]

Although these conclusions cannot be directly extended to other reactions for which the E2C mechanism has been postulated, we believe that if halide ions in dipolar aprotic solvents are sufficiently strong hydrogen nucleophiles to promote an E2H reaction in which C_β— H bond-breaking is a significant cost factor, then there is no compelling need to discard the E2H mechanism when C_β— H bond-breaking is not an important activating process, as in III. In cases where paenecarbonium transition states are favourable, nucleophiles such as RS^- and halide ions in dipolar aprotic solvents often promote more facile elimination than do conventionally strong bases in protic solvents. The low desolvation energies in these systems appear to be an important factor, while thermodynamic basicity must be largely irrelevant because of the small degree of B....H bond formation. Using a different approach to ours, Eck and Bunnett have provided evidence against the E2C mechanism in cases where paenecarbonium transition states are likely to be involved.[1d]

It has been stated that E2C transition states have a well-developed double bond.[2e] Naively visualising the bonding situation at C_α in such a transition state as involving sp^2 hybridization with 4 electrons in a p-type orbital for the B....C....X moiety, it is difficult to see how overlap between this incipient p-type orbital and the incipient p-orbital at C_β containing the best part of 2 electrons, to form a partial π-bond could occur.[12] A charge-separated transition state such as V would overcome this difficulty, but all the available evidence, including that presented here, indicates that negative charge density on C_β is low relative to that found under E2H conditions, whereas this cannot be the case in V.

ACKNOWLEDGEMENT

We thank Professor J. F. Bunnett, University of California, Santa Cruz, for helpful correspondence.

REFERENCES

1.   (a) J. F. Bunnett, Survey of Progress in Chemistry, Vol. 5, P. 53, Academic Press, N.Y. (1969); (b) J. F. Bunnett and E. Baciocchi, J.Org.Chem., 32 11 (1967); (c) idem, ibid., in press; (d) D. Eck and J. F. Bunnett, J.Amer.Chem.Soc., 91 3099 (1969); (e) D. J. McLennan, J.Chem.Soc.(B), 705 (1966); (f) R. J. Anderson, P. Ang, B. D. England, V. H. McCann and D. J. McLennan, Aust.J.Chem., 22 1427 (1969).

2. (a) S. Winstein, <u>Accad. Nazionale di Lincei (Roma), VIII Corso Estivo di Chimica, Chimica Teorica</u>, 327 (1965); (b) N. H. Cromwell, <u>Proc.Chem.Soc.</u>, 252 (1961); (c) A. J. Parker, M. Ruane, G. Biale and S. Winstein, <u>Tetrahedron Letters</u>, 2113 (1968); (d) D. J. Lloyd and A. J. Parker, <u>ibid.</u>, 5183 (1968); (e) D. Cook, A. J. Parker and M. Ruane, <u>ibid.</u>, 5715 (1968); (f) R. Alexander, E. F. C. Ko, A. J. Parker and T. J. Broxton, <u>J.Amer.Chem.Soc.</u>, <u>90</u> 5049 (1968); (g) E. F. C. Ko and A. J. Parker, <u>ibid.</u>, <u>90</u> 6647 (1968).

3. D. N. Kevill, E. D. Weiler and N. H. Cromwell, <u>J.Amer.Chem.Soc.</u>, <u>88</u> 4489 (1966).

4. (a) D. N. Kevill, G. N. Coppens and N. H. Cromwell, <u>J.Amer.Chem.Soc.</u>, <u>86</u> 1553 (1964). (b) D. N. Kevill and J. E. Dorsey, <u>J.Org.Chem.</u>, <u>34</u> 1985 (1969).

5. S. J. Cristol, <u>J.Amer.Chem.Soc.</u>, <u>67</u> 1494 (1945).

6. B. D. England and D. J. McLennan, <u>J.Chem.Soc.(B)</u>, 696 (1966).

7. P. Ang, B. D. England and P. R. Fawcett, unpublished results.

8. G. Baddeley and G. M. Bennett, <u>J.Chem.Soc.</u>, 1819 (1935).

9. C. H. DePuy and C. A. Bishop, <u>J.Amer.Chem.Soc.</u>, <u>82</u> 2532 (1960).

10. L. F. Blackwell, A. Fischer and J. Vaughan, <u>J.Chem.Soc.(B)</u>, 1084 (1967).

11. (a) J. Hine, C. H. Thomas and S. J. Ehrensen, <u>J.Amer.Chem.Soc.</u>, <u>77</u> 3886 (1955); (b) J. Hine, S. J. Ehrensen and W. H. Brader, <u>ibid.</u>, <u>78</u> 2282 (1956).

12. B. Capon and C. W. Rees (eds.), <u>Organic Reaction Mechanisms</u>, 1968, P. 150, Interscience, London, 1969. The validity of using the absence of Bronsted-type correlations as evidence against E2H in halide-promoted eliminations is also questioned here, as elsewhere.[1f]

DDT Resistance in Insects

Rate of Increased Tolerance to Insecticides in the Egyptian Cotton Leafworm, *Spodoptera littoralis* (Boisd.)

By Y. H. ATALLAH[1]

Introduction

Laboratory and field resistance of lepidopterous pests have been reported by several investigators. Resistance in the pink bollworm, *Pectinophora gossypiella* (Saunders), has been reported by CHAPMAN (1960); CHAPMAN and COFFIN (1964), LOWRY and TSAO (1961), LOWRY and BERGER (1965), and LOWRY et al. (1965b). Resistance in the bollworm, *Heliothis zea* (Bobbie), and the tobbacco budworm, *H. virescens* (F.), has been reported by BRAZZEL (1963 and 1964), GRAVES et al. (1963 and 1967), LOWRY et al. (1965a), and LOWRY (1966). Several other cases of resistance and tolerance in lepidopterous pests has been investigated by MCEWEN and SPLITSTOSSER (1964), RABB and GUTHRIE (1964), WOLFENBARGER and LOWRY (1969), and HARDING and DYAR (1970). The Egyptian cotton leafworm, *Spodoptera littoralis* (Boisd.), tolerance and resistance to insecticides had lately received attention by several investigators in Egypt, among these are HASSAN et al (1966), ATALLAH (1971), and HANNA and ATALLAH (1971).

The continuous need of new insecticides for pest control and in some cases the failure of recommended insecticides to acheive the initial control are

---

[1] Laboratory of Plant Protection, National Research Center, Dokki, Cairo, U.A.R.

timely problems. Under field conditions, the insect populations are currently exposed to chemical selection. The selection may be severe when repeatedly using a certain chemical during one season. In this case the insecticide may suppress an insect population, meanwhile those individuals capable of surviving such a severe selection may form a new group of individuals that are capable of tolerating the doses of the selective agent which previously proved lethal to the majority of individuals of the same species. In this case a failure of control is the end result and the need to increase the dose several times to induce the same mortality becomes a necessity. Such populations are termed resistant or tolerant depending upon the rate of increased dosage and the inheritence of the character. This study is an attempt to investigate these points. DDT, endrin, carbaryl, and sumithion®, 0,0-dimethyl 0-(4-nitro-*m*-tolyl) phosphorothioate, has been selected for this study to represent some commonly used insecticides for the control of this pest.

Materials and Methods

A standard laboratory strain of the Egyptian cotton leafworm maintained in the laboratory for over 4 years and fed on caster bean leaves (ATALLAH 1971) was used for this study. The 3rd stage larvae were fed for 24 hrs on caster bean leaves treated with the recommended dose for the control of this pest. At least 2000 larvae were used for each treatment. The larvae were distributed in 1 kg wide mouth glass jars at the rate of 50 larvae/jar. Mortality counts were recorded and the survivors were transferred to new jars and provided with fresh untreated leaves. Daily counts were undertaken for 3 to 4 days as no appreciable change in mortality was observed after this period. The percent mortalities were calculated as daily percents of the total alives of the previous day and as cumulative percents of the original number of individuals subjected to the insecticidal treatment. The survivors were reared and the progeny (F_2) was again treated with the recommended dose, 2, 3, 5, 10, 15, or 20 times that dose as indicated in the results. The F_3 was treated in the same manner. The experiments were aimed at investigating the effect of increasing the recommended dose on the percent mortality of the filial generations.

The tested insecticides were carbaryl, DDT, endrin, and sumithion. Carbaryl was used as 85 % w. p. at the concentration of 0.4 % active ingredient. DDT was used as 25 % e. c. at the concentration of 0.25 % active ingredient. Endrin was used as 19.5 % e. c. at the rate of 0.1 % active ingredient. Sumithion was used as 100 % e. c. at the concentration of 0.44 % active ingredient. The fold increase of the insecticide and the design for treatment differred from one insecticide to the other depending upon the response of the strain to the increased concentration and these are indicated in the results for each insecticide.

Results and Discussion

Carbaryl

Table 1 shows the daily and cumulative percent mortalities resulting from treatment with different concentrations of carbaryl for one generation. The cumulative percent mortality increased with the increase in carbaryl concentration. However, the daily percent mortality at higher concentrations did not increase tremendously, indicating an almost similar latent toxicity in the different concentrations during the F_1.

Table 2 shows the percent mortality during the 2nd generation. Treatment with carbaryl during the 1st generation did not have enough selective power to show a significant difference in the cumulative mortality between

77

Table 1

Daily and cumulative percent mortality of the 3rd stage larvae of the Egyptian cotton leafworm fed for one generation on caster bean leaves treated with carbaryl

	Mortality % Days after treatment					
	1 day		2 days		3 days	
% Carbaryl	Daily	Cumulative	Daily	Cumulative	Daily	Cumulative
0.4	45.5	45.5	27.3	60.4	24.7	70.2
2.0	61.0	61.0	56.7	83.1	69.2	94.7
4.0	69.6	69.6	52.6	85.6	77.8	96.8
6.0	83.0	83.0	51.8	91.8	74.4	97.9
8.0	94.6	94.6	31.5	96.3	59.5	98.5

Table 2

Daily and cumulative percent mortality of the 3rd stage larvae of the Egyptian cotton leafworm fed for 2 generations on caster bean leaves treated with carbaryl

		Mortality % Days after treatment					
% Carbaryl		1 day		2 days		3 days	
F_1	F_2	Daily	Cumulative	Daily	Cumulative	Daily	Cumulative
0.4	0.4	42.1	42.1	7.1	46.2	6.5	49.7
0.4	2.0	58.3	58.3	15.7	64.8	11.5	68.9
0.4	4.0	62.4	62.4	23.1	71.1	37.1	81.8
0.4	6.0	74.0	74.0	23.1	80.0	64.0	92.8
0.4	8.0	82.7	82.7	37.6	89.2	82.4	98.1
8.0	8.0	75.0	75.0	28.3	82.1	62.6	93.3

Table 3

Daily and cumulative percent mortality of the 3rd stage larvae of the Egyptian cotton leafworm fed for 3 or 4 generations on caster bean leaves treated with carbaryl

				Mortality % Days after treatment					
	% Carbaryl			1 day		2 days		3 days	
F_1	F_2	F_3	F_4	Daily	Cumulative	Daily	Cumulative	Daily	Cumulative
0.4	0.4	0.4	—	43.0	43.0	19.5	54.1	13.7	60.4
0.4	0.4	2.0	—	58.3	58.3	24.0	68.3	17.4	73.8
0.4	0.4	4.0	—	65.0	65.0	45.7	81.0	86.8	97.5
0.4	0.4	6.0	—	71.9	71.9	38.4	82.7	89.6	98.2
0.4	0.4	8.0	—	80.5	80.5	38.0	87.9	77.7	97.3
0.4	2.0	6.0	—	79.8	79.8	45.5	89.0	3.7	89.4
0.4	2.0	6.0	8.0	56.0	56.0	50.1	78.1	54.4	90.0
8.0	8.0	8.0	—	25.7	25.7	22.4	42.3	86.1	92.0

the F_1 and F_2 (tables 1 and 2). In the F_2, there was a strong positive relationship between carbaryl concentration and daily mortality during the 2nd and 3rd days indicating a higher latent toxicity at higher concentrations.

Table 3 shows the percent mortality of the 3rd stage larvae during the 3rd and 4th generations of continous selection with carbaryl. The differences in percent mortality did not differ as much as it did during the 1st and 2nd generations. The daily mortality during the 2nd and 3rd days increased at medium concentrations, then decreased at higher concentrations.

Tables 1, 2, and 3 indicates that the upper 3–8 % of the experimental population are resistant to carbaryl, since the higher concentrations did not affect this segment of the population. As a result the mortality never reached 100 %, it almost levelled off when the concentration exceeded a certain limit, i.e. further increase of the concentration did not affect the final mortality. The resistant segment could easily be detected in the 3rd day after treatment. Three generations selection (table 3) shows higher cumulative percent mortalities at similar concentrations during the F_2.

DDT

Table 4 shows the daily and cumulative percent mortality of the 3rd stage larvae of the Egyptian cotton leafworm fed on caster bean leaves treated with DDT at different concentrations for 1, 2, or 3 successive generations. The daily and cumulative mortality increased with the increase in concen-

Table 4

Daily and cumulative percent mortality of the 3rd stage larvae of the Egyptian cotton leafworm fed on caster bean leaves treated with DDT for 1, 2, or 3 generations

			Mortality % Days after treatment							
			1 day		2 days		3 days		4 days	
F_1	% DDT F_2	F_3	Daily	Cum.	Daily	Cum.	Daily	Cum.	Daily	Cum.
0.25	—	—	41.0	41.0	16.0	50.4	11.5	56.1	20.7	65.2
0.5	—	—	62.8	62.8	37.1	76.6	32.5	84.2	33.5	89.5
0.75	—	—	76.3	76.3	52.3	88.7	46.0	93.9	85.2	99.1
0.25	0.25	—	41.1	41.1	14.8	49.8	11.0	55.3	21.5	64.9
0.25	0.5	—	56.4	56.4	28.0	68.6	10.5	71.9	18.9	77.2
0.25	0.25	0.25	49.5	49.5	7.3	53.2	3.4	54.8	7.7	58.3
0.25	0.25	0.5	56.9	56.9	33.3	71.2	43.1	83.6	86.6	97.8

tration. DDT at the concentrations of 0.5 and 0.75 % for one generation yielded 89.5 and 99.1 % mortality after 4 days, most of the survivors failed to pupate or emerged as deformed adults which did not mate or oviposit. Only, the DDT survivors at the concentration of 0.25 % gave adults able to mate and oviposit. This concentration was used to treat the laboratory strain for 2 or 3 successive generations. The data indicates that selection for DDT resistance, if possible in this strain, should be carried out at lower doses. The latent toxicity of DDT when applied for several generations is demonstrated by comparing the cumulative mortality of 0.5 % DDT during the 2nd and 3rd generations.

Endrin

Table 5 shows the toxicity of different rates of endrin to the 3rd stage of the Egyptian cotton leafworm for 1, 2, or 3 successive generations. A considerable increase in mortality was noticed during the 2nd, 3rd, and 4th days following treatment. The data strongly suggests a latent toxicity from one generation to another as a result of successive applications for 2 or 3 generations. Pupation and adult emergence among the survivors of endrin treatment was noticiably high. Endrin at the concentration of 0.5 % for one

Table 5

Daily and cumulative percent mortality of the 3rd stage larvae of the Egyptian cotton leafworm fed on caster bean leaves treated with Endrin for 1, 2, or 3 generations

			Mortality % Days after treatment							
	% Endrin		1 day		2 days		3 days		4 days	
F_1	F_2	F_3	Daily	Cum.	Daily	Cum.	Daily	Cum.	Daily	Cum.
0.1	—	—	49.1	49.1	29.5	64.1	43.7	79.8	50.5	90.0
0.2	—	—	62.7	62.7	33.8	75.3	38.1	84.7	70.0	95.4
0.5	—	—	68.2	68.2	66.7	89.4	70.8	96.9	100	100
0.1	0.1	—	45.0	45.0	46.4	70.5	43.1	83.2	50.6	91.7
0.1	0.2	—	54.5	54.5	58.7	81.2	58.0	92.1	77.0	98.2
0.1	0.5	—	54.5	54.5	62.6	83.0	48.8	91.3	14.9	92.6
0.1	0.1	0.1	46.3	46.3	43.0	69.4	52.3	85.4	33.6	90.3
0.1	0.1	0.2	55.0	55.0	44.2	74.9	52.2	88.0	70.8	96.5
0.1	0.1	0.5	62.4	62.4	43.1	78.6	68.2	93.2	57.4	97.1

generation gave 100 % mortality after 4 days, this might probably be due to the absence of resistant genes in the experimental population, or the unability of the survivors of the previous days to tolerate such high concentrations of endrin. Slight toleration of this pest to endrin could be observed after continous selection for 3 successive generations. The increase in mortality as a result of increased concentrations was higher during the 1st gention than the 2nd and 3rd generations.

Sumithion

Sumithion seems to be the most effective among the tested insecticides in view of its toxicity 1 day after treatment, in addition to the 100 % mortality abtained after 2 days when doubling the original concentration. Table 6

Table 6

Daily and cumulative percent mortality of the 3rd stage larvae of the Egyptian cotton leafworm fed on caster bean leaves treated with Sumithion for 1, 2, or 3 generations

			Mortality % Days after treatment							
	% Sumithion		1 day		2 days		3 days		4 days	
F_1	F_2	F_3	Daily	Cum.	Daily	Cum.	Daily	Cum.	Daily	Cum.
0.44	—	—	96.2	96.2	60.5	98.5	20.0	98.8	8.3	98.9
0.88	—	—	99.1	99.1	100	100				
0.44	0.44	—	90.6	90.6	92.4	99.3	28.6	99.5	60.0	99.8
0.44	0.88	—	98.5	98.5	100	100				
0.44	0.44	0.44	89.8	89.8	84.8	98.5	27.6	98.9	9.1	99.0
0.44	0.44	0.88	98.1	98.1	46.2	99.0	14.3	99.1	33.3	99.4

illustrates the daily and cumulative percent mortality resulting from feeding the 3rd stage larvae with caster bean leaves treated with different concentrations of sumithion for 1, 2, or 3 successive generations. The daily mortality gradualy decreased from the 1st, 2nd, to 3rd days following treatment. No appreciable change in mortality occured after selection for 3 generations. Individuals surviving selection with 0.44 % sumithion for 2 successive generations had some survivors till the 4th day when treated with a doubled concentration during the 3rd generation, while the 1st and 2nd generations had no survivors at the same concentration after 2 days.

Discussion

The difference between the 1st and 4th day mortality when feeding on leaves treated with the recommended concentration for one generation were 24.7, 24.2, 40.9, and 2.7 % for carbaryl, DDT, endrin, and sumithion, respectively. The same differences in mortality when treating for 2 generations were 7.6, 23.8, 46.7 and 9.2 % for the same insecticides respectively, while the same data when treating for 3 generations were 17.4, 8.8, 44.0, and 9.2 %. The acute toxicity of sumithion is the highest among the tested insecticides, however its latent toxicity is the lowest. The experimental population showed a great possibility for resistance to carbaryl. Resistance or tolerance to DDT, endrin, or sumithion in this population is not very likely because of the latent toxicity of DDT and endrin, and the high acute toxicity of sumithion.

References

ATALLAH, Y. H., 1971: Status of carbaryl- and DDT-resistance in laboratory reared Egyptian cotton leafworm. J. Econ. Ent. **64**, 1018–1021.
BRAZZEL, J. R., 1963: Resistance to DDT in *Heliothis virescens*. J. Econ. Ent. **56**, 571–574.
— 1964: DDT resistance in *Heliothis zea*. J. Econ. Entomol. **57**, 455–457.
CHAPMAN, A. J.; COFFIN, L. B., 1964: Pink bollworm resistance to DDT in the Laguna area of Mexico. J. Econ. Ent. **57**, 148–150.
CHAPMAN, R. K., 1960: Status of insecticide resistance in insects attacking vegetable crops. Misc. Publ. Ent. Soc. Amer. **2**, 27–39.
GRAVES, J. B.; CLOWER, D. F.; BRADLEY, JR., J. R., 1967: Resistance of the tobacco budworm to several insecticides in Louisiana. J. Econ. Ent. **60**, 887–888.
GRAVES, J. B.; ROUSSEL, J. S.; PHILLIPS, J. R., 1963: Resistance to some chlorinated hydrocarbon insecticides in the bollworm, *Heliothis zea*. J. Econ. Ent. **56**, 442–444.

HANNA, M. A.; ATALLAH, Y. H., 1971: Penetration and biodegradation of carbaryl in susceptible and resistant strains of the Egyptian cotton leafworm. J. Econ. Ent. **64**, 1391–1394.

HARDING, J. A.; DYAR, R. C., 1970: Resistance induced in European corn borers in the laboratory by exposing successive generations to DDT, diazinon, or carbaryl. J. Econ. Ent. **63**, 250–253.

HASSAN, S. M.; ZAKI, M. M.; ABO-ELGHAR, M. R.; HANNA, M. A., 1966: Some factors affecting the susceptibility of the cotton leafworm, *Prodenia litura* (F.) to insecticides in biological tests. Bull. Ent. Soc. Egypte, Econ. Ser. **I**, 127–142.

LOWRY, W. L., 1966: Bollworm and tobacco budworm resistance to some insecticides in the Lower Rio Grande Valley in 1964. J. Econ. Ent. **59**, 479–480.

LOWRY, W. L.; BERGER, R. S., 1965: Investigations of pink bollworm resistance to DDT in Mexico and the United States. J. Econ. Ent. **58**, 590–591.

LOWRY, W. L.; TSAO, C. H., 1961: Incidence of pink bollworm resistance to DDT. J. Econ. Ent. **54**, 1209–1211.

LOWRY, W. L.; McGARR, R. L.; ROBERTSON, O. T.; BERGER, R. S.; GRAHAM, H. M., 1965a: Bollworm and tobacco budworm resistance to several insecticides in the Lower Rio Grande Valley of Texas. J. Econ. Ent. **58**, 732–734.

LOWRY, W. L.; OUYE, M. T.; BERGER, R. S., 1965b: Rate of increase in resistance to DDT in pink bollworm adults. J. Econ. Ent. **58**, 781–782.

McEWEN, F. L.; SPLITSTOSSER, C. M., 1964: A genetic factor controlling color and its association with DDT sensitivity in the cabbage looper. J. Econ. Ent. **60**, 568–573.

RABB, R. L.; GUTHRIE, F. E., 1964: Resistance of tocacco hornworms to certain insecticides in North Carolina. J. Econ. Ent. **57**, 995–996.

WOLFENBARGER, D. A.; LOWRY, W. L., 1969: Toxicity of DDT and related compounds to certain lepidopteran cotton insects. J. Econ. Ent. **62**, 432–435.

Status of Carbaryl and DDT Resistance in Laboratory-Reared Egyptian Cotton Leafworm[1]

Yousef H. Atallah

Numerous cases of field resistance in lepidopterous insects to insecticides have been reported (Chapman 1960; Lowry and Tsao 1961; Graves et al. 1963; Brazzel 1963, 1964; Chapman and Coffin 1964; McEwen and Splitstosser 1964; Rabb and Guthrie 1964; Lowry and Berger 1965; Lowry et al. 1965a; Lowry 1966; Graves et al. 1967; Wolfenbarger and Lowry 1969). These workers reported resistance in the bollworm, *Heliothis zea* (Boddie); the tobacco budworm, *H. virescens* (F.); the tobacco hornworm, *Manduca sexta* (Johannson); the pink bollworm, *Pectinophora gossypiella* (Saunders); and the cabbage looper, *Trichoplusia ni* (Hübner).

Laboratory studies on resistance in lepidopterous insects are comparatively few. Lindquist and Dahm (1956) found that resistance of the European corn borer, *Ostrinia nubilalis* (Hübner), to DDT was primarily due to slow absorption (15% of the applied dose in 120 hr). Adkisson and Wellso (1962) reported that sublethal doses of DDT on the pink bollworm lowered the incidence of mating and adult longevity. Lowry et al. (1965b) found that DDT treatment of the adult pink bollworm for 21 generations had no effect on the weight. However there was an 18-fold increase in resistance. Sublethal doses of endrin,

---

[1] Lepidoptera: Noctuidae.

83

azinphosethyl, or azinphosmethyl reduced larval weight, adult longevity, and number of eggs per adult of the bollworm and the tobacco budworm, but the treatment had no effect on egg viability (Chauthani and Adkisson 1966). Atallah and Newsom (1966) treated adults of a lady beetle, *Coleomegilla maculata* De Geer, with sublethal doses and found that DDT increased oviposition, endrin had no effect, while toxaphene prevented oviposition. Harding and Dyar (1970) selected for resistance in the European corn borer in the laboratory by exposing successive generations to DDT, diazinon, or carbaryl. The selection resulted in an increased tolerance to the test chemical. However, by the 12th generation the reproduction of the borer was reduced as well as the weights of larvae and pupae of both sexes. Hanna (1970[*]) found that sublethal doses of DDT, endrin, carbaryl, or Sumithion® (*O,O*-dimethyl *O*-(4-nitro-*m*-tolyl) phosphorothioate) affected differently the reproductive and survival potentials of the Egyptian cotton leafworm, *Spodoptera littoralis* (Boisduval), and the effect lasted for several generations (sometimes up to 12 generations). This paper reports the toxicity of carbaryl and DDT to laboratory-reared colonies of *S. littoralis* collected from different localities in the U.A.R. and the development of resistance and susceptibility in these colonies.

MATERIALS AND METHODS.—Egyptian cotton leafworms were collected from 4 localities in the U.A.R. during the autumn of 1966. Four laboratory colonies were established, labeled 1, 2, 3, and 4. Castor bean leaves were used for feeding the larvae, since they were available the year around and kept fresh for a comparatively long period. The larvae were allowed to pupate in sawdust. The pupae were sexed and held in glass jars until adults emerged. Each 2 pairs of moths were held in a 1-kg glass jar and supplied with a 10% honey solution for food and a small branch of tafla, *Nerium oleander,* for oviposition. The eggs were collected daily and held in 225-ml glasses until they hatched.

Three methods of selection of resistant individuals were carried out simultaneously. The 1st method was by topical treatment on the notum of 3rd-stage larvae; at least 5000 individuals were treated each generation. A microsyringe fitted to a micrometer was used for the delivery of 1 μliter of acetone containing the required dose. The LD_{80} to LD_{95} rates were used for selection. The survivors were reared and allowed to reproduce. The log dose-probit (ld-p) lines were established for the survivors, and reselection was carried out in the filial generations when the number of individuals was large enough, otherwise genera-

[*] M. A. Hanna. 1970. New synthetic pesticides and their toxicity to the cotton leafworm, *Spodoptera littoralis* (Boisduval). Ph.D. dissertation, Faculty of Agriculture Library, Ain Shams University, Cairo, U.A.R.

84

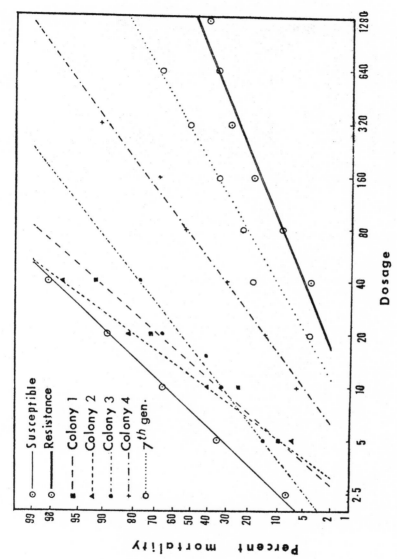

Fig. 1.—Log dose-probit lines for carbaryl-treated 3rd-stage larvae of the Egyptian cotton leafworm.

tions having a small number of individuals were not treated, or treated with the LD_{10} to LD_{50} rates. Topical application of large doses of carbaryl presented a problem. Solubility of carbaryl in acetone did not allow a dose over 95 $\mu g/\mu$liter (and crystallization of the chemical in the needle was troublesome. Also large doses crystallized quickly on the notum and fell off, giving erratic results. To solve this problem a simple technique was devised by dipping the 3rd-stage larvae in a solution of carbaryl in acetone:water mixture (4:1 by volume). Mortalities were recorded 24 hr later. Comparison of the ld-p lines from dipping and topical application on the notum of 3rd-stage larvae proved that the 2 lines were parallel. The concentration of carbaryl in the topical application solution was 19-fold that in the dipping solution for the same mortality. The dipping technique was used whenever the dose exceeded 100 μg/larva (6000 $\mu g/g$).

The 2nd method involved feeding large numbers (at least 10,000 individuals/treated generation) of 3rd-stage larvae on treated leaves. The proper amount of the formulated insecticide was added to water to obtain the desired concentration. Fresh castor-bean leaves were dipped in the proper concentration, then air dried. The dose on the treated leaves was designed to allow about 80% mortality after 24 hr. In this method, 24-hr survivors usually had higher mortality during their development than survivors of the 1st method.

The 3rd method was by rearing the progeny of each pair alone, then testing the progeny for resistance. If the response of tested individuals was homogeneous and their LD_{50} exceeded (at least double) that of the mother colony, their siblings were reared; otherwise the progeny was discarded. The susceptible strain was established through rearing siblings of susceptible individuals.

During this work the 3 methods were combined in selection of the resistant individuals from the established colonies (i.e. Colonies 1, 2, 3, and 4).

The survivors of the 3 methods of selection were pooled and reared as 1 colony, then their progeny was subjected to reselection by the 3 methods generation after generation. Selection for resistance using this combination proved satisfactory. Selective doses were always increased from one generation to the other, depending on the response of the animal to selection.

RESULTS AND DISCUSSION.—Field-collected colonies of the Egyptian cotton leafworm from 4 localities in the UAR showed varying responses to DDT or carbaryl treatments. The difference in susceptibility was probably due to selection as a result of the extensive use of these 2 chemicals on the Egyptian cotton leafworm populations, or cross resistance with other commonly used insecticides.

Carbaryl.—Ld-p lines were established for each

colony (Fig. 1). Colonies showing heterogeneity in their response and comparatively large LD_{50} values for carbaryl were chosen for selection toward resistance. In breeding for carbaryl resistance, colonies 3 and 4 were used. The average LD_{50} for the 2 colonies was about 3333 $\mu g/g$. The survivors after 7 generations of continuous selection by the 3 methods simultaneously had an LD_{50} of 25,000 $\mu g/g$, and after 15 generations it reached 100,000 $\mu g/g$ (Resistance, Fig. 1). Selection by topical application was discontinued because of the extensive amount of carbaryl to be used on 1 larva and difficulties in applying this dose. Dipping the larvae in a solution of carbaryl in acetone: water mixture (4:1 by volume) substituted the topical application technique. The final ld-p line for the carbaryl-resistant strain (CR-Strain) tended to level off after the LD_{30}, therefore the LD_{50} was determined by extrapolation. A susceptible strain was established through rearing siblings of susceptible individuals for 8 generations. The carbaryl-susceptible strain (CS-Strain) had an LD_{50} of 500 $\mu g/g$ and a steep ld-p line indicating a homogeneous response in the strain to carbaryl (Fig. 1). The CR-Strain was 200-fold more resistant than CS-Strain and 30-fold more resistant than the parental populations. Rearing the CS-Strain in the laboratory for 26 generations indicated no apparent change in its response to carbaryl, meanwhile rearing CR-Strain away from insecticidal treatment for 26 generations indicated a slight loss in its resistance to carbaryl. The CR-Strain became 24-fold instead of 30-fold resistant compared with the parental population and 160-fold instead of 200-fold compared with the CS-Strain.

DDT.—Ld-p lines were established for each of the 4 field-collected colonies (Fig. 2). The variations in response to DDT treatment among the colonies were considerably less than in case of carbaryl. In breeding for DDT resistance, the 4 colonies were used. Selection for DDT resistance was a problem, since treatment of the colonies with DDT, as low as the LD_{40}, resulted in deterioration of the colonies. Adult females developing from treated larvae had a lower reproductive potential, and loss of egg viability was common. This problem is under investigation. Selection at the level of LD_{30} for 6 generations resulted in no apparent change in response to DDT; the same happened when breeding siblings of tolerant individuals. After 18 generations of alternative selections (selection every other generation) at the level of LD_{30} and extreme care in handling the colony, tolerance to DDT reached 6-fold compared with the parental populations (Fig. 2). A DDT-susceptible strain was established through rearing siblings of susceptible individuals for 8 generations. The LD_{50} values were 100, 200, and 1200 $\mu g/g$ for the susceptible strain, parental populations, and tolerant strain. Rearing the DDT-susceptible strain in the laboratory for 24 generations indicated no apparent change in its response to DDT,

87

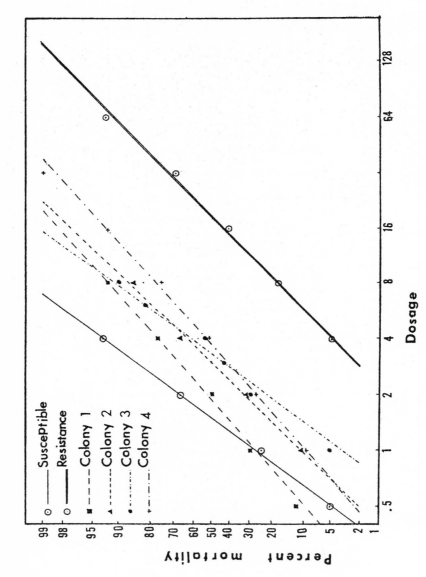

Fig. 2.—Log dose-probit lines for DDT-treated 3rd-stage larvae of the Egyptian cotton leafworm.

and its ld-p line was steep, indicating a homogeneous population. When the DDT-tolerant strain was reared away from any insecticidal pressure for 24 generations the LD_{50} decreased to about 560 $\mu g/g$.

Maintaining the established strains, especially the CR-Strain and the DDT-tolerant strain under laboratory conditions was very difficult because of diseases attacking this pest when reared in the laboratory. Decontamination of the rearing room and equipment was essential during this work.

REFERENCES CITED

Adkisson, P. L., and S. G. Wellso. 1962. Effect of DDT poisoning on the longevity and fecundity of the pink bollworm. J. Econ. Entomol. 55: 842–5.

Atallah, Y. H., and L. D. Newsom. 1966. Ecological and nutritional studies on *Coleomegilla maculata* De Geer (Coleoptera:Coccinellidae). III. The effect of DDT, toxaphene, and endrin on the reproductive and survival potentials. Ibid. 59: 1181–7.

Brazzel, J. R. 1963. Resistance to DDT in *Heliothis virescens*. Ibid. 56: 571–4.

1964. DDT resistance in *Heliothis zea*. Ibid. 57: 455–7.

Chapman, A. J., and L. B. Coffin. 1964. Pink bollworm resistance to DDT in the Laguna area of Mexico. J. Econ. Entomol. 57: 148–50.

Chapman, R. K. 1960. Status of insecticide resistance in insects attacking vegetable crops. Misc. Publ. Entomol. Soc. Amer. 2: 27–39.

Chauthani, A. R., and P. L. Adkisson. 1966. Effects of sublethal doses of certain insecticides on eggs, larvae, and adults of two species of *Heliothis*. Ibid. 59: 1070–4.

Graves, J. B., D. F. Clower, and J. R. Bradley, Jr. 1967. Resistance of the tobacco budworm to several insecticides in Louisiana. Ibid. 60: 887–8.

Graves, J. B., J. S. Roussel, and J. R. Phillips. 1963. Resistance to some chlorinated hydrocarbon insecticides in the bollworm, *Heliothis zea*. Ibid. 56: 442–4.

Harding, J. A., and R. C. Dyar. 1970. Resistance induced in European corn borers in the laboratory by exposing successive generations to DDT, diazinon, or carbaryl. Ibid. 63: 250–3.

Lindquist, D. A., and P. A. Dahm. 1956. Metabolites of radioactive DDT by the Madeira roach and European corn borer. Ibid. 49: 579–84.

Lowry, W. L. 1966. Bollworm and tobacco budworm resistance to some insecticides in the Lower Rio Grande Valley in 1964. Ibid. 59: 479–80.

Lowry, W. L., and C. H. Tsao. 1961. Incidence of pink bollworm resistance to DDT. Ibid. 54: 1209–11.

Lowry, W. L., and R. S. Berger. 1965. Investigations of pink bollworm resistance to DDT in Mexico and The United States. Ibid. 58: 590–1.

Lowry, W. L., R. L. McGarr, O. T. Robertson, R. S. Berger, and H. M. Graham. 1965a. Bollworm and tobacco budworm resistance to several insecticides in the Lower Rio Grande Valley of Texas. Ibid. 58: 732–4.

Lowry, W. L., M. T. Ouye, and R. S. Berger. 1965b. Rate of increase in resistance to DDT in pink bollworm adults. Ibid. 58: 781–2.

McEwen, F. L., and C. M. Splitstosser. 1964. A genetic factor controlling color and its association with DDT sensitivity in the cabbage looper. Ibid. 57: 568–73.

Rabb, R. L., and F. E. Guthrie. 1964. Resistance of tobacco hornworms to certain insecticides in North Carolina. Ibid. 57: 995–6.

Wolfenbarger, D. A., and W. L. Lowry. 1969. Toxicity of DDT and related compounds to certain lepidopteran insects. Ibid. 62: 432–5.

Species Specificity Among Oriental Fruit Flies,[1] Melon Flies,[1] and Mediterranean Fruit Flies[1] in Susceptibility to Insecticides at Several Loci[2]

Irving Keiser, Esther L. Schneider, and Isao Tomikawa

In the course of laboratory assessment of several hundred candidate insecticides against oriental fruit flies, *Dacus dorsalis* Hendel; melon flies, *D. cucurbitae* Coquillett; and Mediterranean fruit flies, *Ceratitis capitata* (Wiedemann), we routinely applied micro-amounts of the chemicals in acetone solution to the thoracic mesonotum, an accepted standard procedure used by March and Metcalf (1949) and others. We found it easier, however, to apply this small dosage to the abdominal mid-venter than to the thoracic mesonotum, since the former region is not so sharply limited as the latter. While we assumed at first that there would be no differences in subsequent toxicities from either method of application, nevertheless we ran comparative tests with oriental fruit flies to make sure this supposition was correct. We found the reverse to be true, that identical quantities of the

---

[1] Diptera: Tephritidae.

Mention of a proprietary product does not necessarily imply endorsement by the USDA. This is a report of research results and not a recommendation of any of the materials tested.

same insecticide applied to the abdominal mid-venter resulted in mortalities that were very different from those applied to the thoracic mesonotum.

This report details the results of extensive tests made in 1957 as a result of the aforementioned findings, employing all 3 species of Hawaiian tephritids and treating 6 different body regions as well as the wings with several different insecticides. Also, a repetition and expansion of these studies was made about 10 years later, in 1967, when we found, inadvertently, while studying the effect of insecticides on flies sexually sterilized with tepa or radiation (Keiser and Schneider 1969), that melon flies had developed complete resistance to DDT when applied to the thoracic mesonotum sometime during this 10-year interval.

O'Kane et al. (1933) studied controlled applications, using a platinum needle dipped in 95% nicotine and applied to various regions of last stadial larvae of the yellow mealworm, *Tenebrio molitor* L. They noted that applications to the mouth parts or spiracles or anal opening did not always give maximum reaction. Hockenyos and Lilly (1932) made hypodermic injections of nicotine sulphate into the larvae of the white-lined sphinx, *Celerio lineata* (F.), and found that the speed of paralyzation was in direct proportion to the distance from the point of injection to the head. Fisher (1952) showed that the effectiveness of crystalline DDT against the house fly, *Musca domestica* L., increased as loci of application approached the body or head. Ahmed and Gardiner (1968) studied variation in toxicity of malathion when applied to different body regions of the desert locust, *Schistocerca gregaria* (Forskal). We report comparative studies conducted with several related species and with different insecticides simultaneously.

METHODS.—We prepared geometric dilutions of acetone solutions of technical DDT, methoxychlor, and malathion ranging from 0.025% to 0.001%. In the 1957 studies, 1 μliter of solution was applied by means of a microdevice described by Roan and Maeda (1953) to the thoracic mesonotum of each of 20 adults of mixed sexes at each dosage level, while the insects were immobilized with CO_2 (Williams 1946). In the 1967 studies, we used a Model M microapplicator® (Instrument Specialities Co.), also calibrated to deliver 1 μliter of solution. Mortality counts were made after 24 hr, dose-mortality curves were computed (Daum 1970), and LD_{50} and LD_{95} levels were noted. A minimum of 7 dosages was used for each insecticide, and all tests were replicated 2–3 times.

Application loci were: thoracic mesonotum, thoracic mesosternum, abdominal mid-dorsum, abdominal mid-venter, verto-occipual region, and oral region. We placed the μliter of solution in the center of each body region, and no spread was noted to adjacent body regions. However, the thoracic mesosternal application involved some contact with the coxae. Treatment on the wings with DDT and methoxychlor produced no toxicity to any of the 3 species, and only very low toxicity with malathion. Hence, this locus was not investigated further in these comparative studies.

In 1957, we applied DDT to all 3 species, methoxychlor to oriental fruit flies and Mediterranean fruit flies, and malathion to Mediterranean fruit flies only; in 1967, we tested all 3 species with all 3 insecticides on the same 6 body regions.

All of these studies were made with laboratory-reared fruit flies. In tests with wild melon flies, infested *Momordica* sp. was collected on the Island of Hawaii, and the larvae were allowed to pupate and adults to emerge in Honolulu. During adulthood, they received the same diet as our laboratory strain, namely sugar, water, and hydrolyzed protein. These were treated topically with DDT or malathion when they were 12 and 21 days old, along with laboratory-reared melon flies of the same age. Because of a shortage of adults of the wild strain, we treated only 2 body regions of these melon flies, namely the thoracic mesonotum and thoracic mesosternum.

RESULTS.—*Oriental Fruit Fies.*—DDT.—Table 1 shows that both in the 1957 and 1967 studies, and at the LD_{50} and LD_{95} levels, DDT was least toxic to this species when application was made on the thoracic mesonotum and most toxic when applied to the oral region.

Methoxychlor.—The results with methoxychlor were very similar to those with DDT. In both 1957 and 1967, treatments applied to the thoracic mesonotum were least effective; those on the oral region were most effective. At the LD_{95}, the differences were even more pronounced than at the LD_{50}.

Malathion.—In the 1957 studies with oriental fruit flies, malathion was tested on the thoracic mesonotum only, but in 1967, all of the 6 body regions were treated (Table 1). At the LD_{95}, malathion applied to the thoracic mesonotum was twice as effective as when applied to the abdominal mid-dorsum or abdominal mid-venter; with DDT the opposite was observed, application to either of the 2 last-mentioned regions being 25 times more effective than application to the thoracic mesonotum. With malathion at the LD_{95}, treatments to the oral region were only 2 times more effective than to the thoracic mesonotum, but with DDT and methoxychlor, treatments to the oral region were 32–38 times more effective than to the thoracic mesonotum.

Melon Flies.—DDT.—The 1957 results with DDT against melon flies (Table 2) were similar to those with this insecticide against oriental fruit flies (Table 1), namely that treatments applied to the thoracic mesonotum were least effective, and those to the oral region were most effective of the 6 body regions tested. In 1967, however, melon flies had built up complete resistance to DDT when applied to the thoracic mesonotum (Table 2). Also, although 50%

93

Table 1.—Toxicity to adult oriental fruit flies of insecticides when applied topically to different body regions.

Application loci	Micrograms toxicant per fly to achieve[a]			
	LD-$_{50}$ in—		LD-$_{95}$ in—	
	1957–9	1967	1957–9	1967
DDT				
Thoracic mesonotum	1.57 a	1.64 a	8.20 a	24.0 a
Thoracic mesosternum	0.44 c	0.31 c	1.42 c	0.63 bc
Abdominal mid-dorsum	.80 b	.49 b	2.62 b	1.00 bc
Abdominal mid-venter	.48 c	.43 b	1.67 bc	.89 bc
Verto-occiputal region	.47 c	.55 b	.93 cd	1.39 b
Oral region	.29 d	.31 c	.64 d	.75 bc
Methoxychlor				
Thoracic mesonotum	3.53 a	6.43 a	35.6 a	44.1 a
Thoracic mesosternum	.69 c	.62 d	1.82 c	1.77 cd
Abdominal mid-dorsum	1.23 b	1.46 b	3.12 bc	3.75 b
Abdominal mid-venter	1.37 b	1.10 bc	3.89 b	2.61 bc
Verto-occiputal region	1.03 b	.97 c	4.20 b	1.93 c
Oral region	.47 d	.52 d	1.20 c	1.16 d
Malathion				
Thoracic mesonotum	.051	.033 a	.097	.089 b
Thoracic mesosternum		.016 b		.033 c
Abdominal mid-dorsum		.035 a		.15 ab
Abdominal mid-venter		.044 a		.18 a
Verto-occiputal region		.014 b		.034 c
Oral region		.018 b		.042 c

[a] LD values for insecticides followed by the same letter in a column are not significantly different at the 5% level.

melon fly mortalities were achieved when DDT was applied to the other body regions, in almost all instances 95% mortalities could not be obtained.

Methoxychlor.—In 1957, methoxychlor was tested against melon flies on the thoracic mesonotum only, and in 1967, all 6 body regions were treated. As with DDT, melon flies showed complete resistance to

Table 2.—Toxicity to adult melon flies of insecticides when applied topically to different body regions.

| | Micrograms toxicant per fly to achieve[a] | | | |
| | LD-50 in— | | LD-95 in— | |
Application loci	1957-9	1967	1957-9	1967
DDT				
Thoracic mesonotum	17.4 a	>256.0[b]	78.5 a	>256.0[b]
Thoracic mesosternum	2.39 cd	7.31	12.4 cd	92.4
Abdominal mid-dorsum	5.19 b	6.40	35.7 ab	120.0
Abdominal mid-venter	2.82 c	27.5	20.9 bc	>256.0[b]
Verto-occiputal region	3.10 c	16.0	20.7 bc	>256.0[b]
Oral region	1.94 d	5.0	6.70 d	>256.0[b]
Methoxychlor				
Thoracic mesonotum	8.67	>256.0[b]	28.5	>256.0[b]
Thoracic mesosternum		2,052 b		11.1 b
Abdominal mid-dorsum		33.8		>256.0[b]
Abdominal mid-venter		18.2		>256.0[b]
Verto-occiputal region		2.79 a		34.4 a
Oral region		1.24 c		17.9 ab
Malathion				
Thoracic mesonotum	0.066	0.066 c	0.12	0.23 b
Thoracic mesosternum		.025 d		.049,d
Abdominal mid-dorsum		11.h		.34 b
Abdominal mid-venter		.17 a		.75 a
Verto-occiputal region		.038 c		.099 c
Oral region		.049 c		.26 b

[a] LD values for insecticides followed by the same letter in a column are not significantly different at the 5% level.
[b] Tested up to 256.0 μg toxicant/fly. Mortalities too low for valid regression.

methoxychlor in 1967 when it was applied to the thoracic mesonotum (Table 2).

Malathion.—Malathion was tested against melon flies in 1957 on the thoracic mesonotum only, and on all 6 body regions in 1967 (Table 2). As with oriental fruit flies (Table 1), the results with malathion against melon flies were entirely different from those with DDT and methoxychlor, namely, the thoracic mesonotum was not the least effective treatment body region. In fact, 1967 treatments on the abdominal mid-dorsum and on the abdominal mid-venter with malathion were 2–3 times less effective than those applied to the thoracic mesonotum, although they were at least 8–40 times more effective when DDT or methoxychlor were the insecticides applied.

Mediterranean Fruit Flies.—DDT.—Mediterranean fruit flies reacted entirely differently to DDT than oriental or melon flies. As noted in Table 3, in the 1957 tests and at the LD$_{50}$, treatments on Mediterranean fruit fly thoracic mesonota were the most effective of all treatment sites although they were least effective when applied to this region for oriental and melon flies (Tables 1 and 2). Also, at the LD$_{95}$, application to the thoracic mesonotum of Mediterranean fruit flies was one of the most effective, and to the oral region the least effective of the 6 body regions treated—again a complete reversal of results with oriental and melon flies. In the 1967 studies, toxicities at the LD$_{50}$ for all regions were alike. This similarity was also noted at the LD$_{95}$, except for applications on the abdominal mid-dorsum and mid-venter.

Methoxychlor.—As with DDT, Mediterranean fruit flies reacted entirely differently than oriental fruit flies or melon flies to methoxychlor. In the 1957 studies, the thoracic mesonotum was one of the most susceptible regions (Table 3), while the abdominal mid-venter and verto-occiputal regions were among the least susceptible. Again in 1967, and as with DDT, toxicities from methoxychlor at the LD$_{50}$ for all regions were mostly alike, and also at the LD$_{95}$ except for topical applications on the abdominal mid-dorsum and ventro-occiputal regions.

Malathion.—Results with malathion against Mediterranean fruit flies were similar in many respects to those with DDT and methoxychlor, for both 1957 and 1967 (Table 3), as compared with the very dissimilar results with malathion vs. DDT or methoxychlor against oriental and melon flies (see Tables 1 and 2).

Wild vs. Laboratory-Reared Melon Flies.—As recorded in Table 4, the wild strain of melon flies also was highly resistant to topical applications of DDT to the thoracic mesonotum. The results of the thoracic mesonotal treatments of wild melon flies were markedly similar to those of the laboratory-reared melon flies. Again, as with the laboratory strain, the wild melon flies were not resistant to malathion

Table 3.—Toxicity to adult Mediterranean fruit flies of insecticides when applied topically to different body regions.

| | Micrograms toxicant per fly to achieve[a] | | | |
| | LD_{50} in— | | LD_{95} in— | |
Application loci	1957-9	1967	1957-9	1967
DDT				
Thoracic mesonotum	0.16 c	0.23 b	1.26 b	0.88 b
Thoracic mesosternum	.21 bc	.19 b	1.30 b	.61 b
Abdominal mid-dorsum	.27 b	.34 a	1.17 b	1.85 a
Abdominal mid-venter	.23 bc	.25 ab	1.19 b	1.18 ab
Verto-occiputal region	.96 a	.24 b	4.45 a	.66 b
Oral region	.78 a	.20 b	4.25 a	.71 b
Methoxychlor				
Thoracic mesonotum	.32 c	.15 b	0.95 c	.52 b
Thoracic mesosternum	.22 d	.20 b	.65 c	.59 b
Abdominal mid-dorsum	.31 c	.27 ab	.93 c	1.70 ab
Abdominal mid-venter	1.29 a	.18 b	4.37 a	.58 b
Verto-occiputal region	1.01 a	.30 a	3.45 a	2.16 a
Oral region	.63 b	.18 b	1.67 b	.97 b
Malathion				
Thoracic mesonotum	.0040 bc	.0029 bcd	.010 b	.045 a
Thoracic mesosternum	.0030 c	.0016 e	.0069 b	.0065 b
Abdominal mid-dorsum	.0034 bc	.0052 ac	.010 b	.012 b
Abdominal mid-venter	.0043 b	.0042 ab	.012 b	.021 ab
Verto-occiputal region	.0093 a	.0030 bd	.064 a	.0079 b
Oral region	.0105 a	.0024 de	.057 a	.18 ab

[a] LD values for insecticides followed by the same letter in a column are not significantly different at the 5% level.

97

(Table 4).

DISCUSSION AND CONCLUSIONS. — We are presented with several interesting phenomena as a result of these investigations. In the first place, oriental and melon fly toxicities from treatments with DDT or methoxychlor were different from those with malathion when applied to identical body regions. For example, DDT or methoxychlor was less toxic to oriental fruit flies or melon flies when applied to the thoracic mesonotum than when applied to the abdominal mid-dorsum or abdominal mid-venter. With malathion, however, the reverse was true; this insecticide was more toxic to oriental fruit flies or melon flies when applied to the thoracic mesonotum than when applied to the abdominal mid-dorsum or abdominal mid-venter. Secondly, identical treatments applied to identical body regions of Mediterranean fruit flies resulted in different toxicity relationships as compared with oriental or melon flies. For example, when oriental fruit flies or melon flies were treated with DDT or methoxychlor on the thoracic mesonotum, toxicities were the lowest of all 6 body regions. However, when Mediterranean fruit flies were treated with DDT or methoxychlor on the thoracic mesonotum, toxicities were among the highest of all body regions.

In almost every study in 1957 and in 1967, the thoracic mesonota of oriental fruit flies and melon flies were the least sensitive, and the thoracic mesosternal and oral regions were the most sensitive to topical applications of DDT and methoxychlor. However, with applications of malathion, differences in susceptibility of different body regions of oriental fruit flies and melon flies were not so pronounced.

The information included in Table 2 showing buildup of resistance to DDT and methoxychlor by the laboratory strain of the melon fly was developed because of a fortunate set of circumstances; if we had not tested tepa-chemosterilized, irradiated, and non-sterilized melon flies with DDT and malathion in 1967, we would never have run the 1967 tests. Evidently, there was a natural and as yet unexplained buildup in resistance of melon flies to DDT and methoxychlor some time during the period from 1957 to 1967, in the laboratory strain. The wild population of melon flies was resistant to DDT by 1969, but the inception of this resistance is unknown.

Table 2 also demonstrates the importance of the LD_{95} in evaluating the development of resistance in melon flies. The LD_{50} has generally been considered by investigators to be the more reliable value, since at this point variation is the least, i.e., the confidence interval is the narrowest. For example, as shown in Table 2, the LD_{50} from treatment at the oral region was 1.94 μg toxicant/fly in 1957 and 5.0 μg toxicant/fly in 1967, a 2½-fold increase. Yet the LD_{95} went from 6.70 μg to more than 256.0 μg (LD_{95} in 1967 was not reached, even when DDT-saturated acetone was

98

Table 4.—Comparative toxicity of insecticides to wild and laboratory populations of melon flies, as evaluated by laboratory topical treatments. 1967.

Application loci	Micrograms toxicant per fly to achieve[a]							
	DDT				Malathion			
	LD_{50}		LD_{95}		LD_{50}		LD_{95}	
	Wild	Lab	Wild	Lab	Wild	Lab	Wild	Lab
Thoracic mesonotum	>128.0[b]	>128.0[b]	>128.0[b]	>128.0[b]	0.056 a	0.094 a	0.15 a	0.29 a
Thoracic mesosternum	4.33	4.88	64.4	47.3	.015 b	.016 b	.030 b	.037 b

[a] LD values for insecticides followed by the same letter in a column are not significantly different at the 5% level.
[b] Tested up to 128.0 μg toxicant/fly. Mortalities too low for valid regression.

99

used), or at least 38-fold. When resistance was present, the regression curves became less steep.

Tables 1, and 2, and 3 show data from 1957 and 1967 when we treated all 3 species with malathion applied to the thoracic mesonotum. The similar susceptibilities for both periods indicated no buildup in resistance to malathion during this 10-year interval for any of the 3 species of Hawaiian tephritids.

We are aware that penetration of insecticide-acetone solution into the body of the insect may be different at the various sites treated. We are aware also that the actual resistance or susceptibility to an insecticide does not necessarily occur at the treatment site. Nevertheless, during these studies we noted several phenomena that require explanation and may lead other toxicologists to test additional species of insects with different insecticides and at different treatment sites. The phenomena needing elucidation are that (1) oriental fruit flies and melon flies showed similar susceptibilities to topical applications of DDT and methoxychlor at the same body regions, but entirely different susceptibilities at those body regions to applications of malathion; (2) a closely related species, Mediterranean fruit fly, reacted entirely differently than oriental fruit flies or melon flies, in some respects oppositely, to topical applications of DDT and methoxychlor; (3) melon flies developed complete resistance in that region (thoracic mesonotum) which, as would be expected, was least susceptible to topical application prior to development of such resistance.

REFERENCES CITED

Ahmed, H., and B. G. Gardiner. 1968. Variation in toxicity of malathion when applied to certain body regions of Schistocerca gregaria (Forsk.). Bull. Entomol. Res. 57: 651–9.

Daum, R. J. 1970. Revision of two computer programs for probit analysis. Bull. Entomol. Soc. Amer. 16: 10–15.

Fisher, R. W. 1952. The importance of the locus of application on the effectiveness of DDT for the house fly Musca domestica L. Can. J. Zool. 30: 254–66.

Hockenyos, G. L., and J. Lilly. 1932. Toxicity studies by hypodermic injection of Celerio lineata larvae. J. Econ. Entomol. 25: 253–60.

Keiser, I., and E. L. Schneider. 1969. Longevity, resistance to deprivation of food and water, and susceptibility to malathion and DDT of oriental fruit flies, melon flies, and Mediterranean fruit flies sexually sterilized with tepa or radiation. Ibid. 62: 663–7.

March, R. B., and R. L. Metcalf. 1949. Laboratory and field studies of DDT-resistant house flies in southern California. Calif. Dep. Agr. Bull. 38: 93–101.

O'Kane, W. C., G. L. Walker, H. G. Guy, and O. J. Smith. 1933. Studies of contact insecticides—VI. 1. Reactions of certain insects to controlled applications of various concentrated chemicals. 2. A new technique for initial appraisal of proposed contact insecticides. New Hampshire Agr. Exp. Sta. Tech. Bull. 54. 23 p.

Roan, C. C., and S. Maeda. 1953. A microdevice for rapid application of toxicants to individual insects. USDA, Agr. Res. Serv., Bur. Entomol. Plant Quar., ET-306. 3 p.

Williams, C. M. 1946. Continuous anaesthesia for insects. Science 103: 57.

Changes in Glucose Metabolism Associated with Resistance to DDT and   Dieldrin in the House Fly[1,2,3]

FREDERICK W. PLAPP, JR.

Much progress has been made in identifying biochemical and genetic mechanisms responsible for insect resistance to insecticides, particularly when resistance involves an increase in ability to detoxify insecticides. However, some types of resistance are apparently independent of detoxication, and their nature remains unknown. Examples are resistance to dieldrin and other cyclodiene insecticides and at least 1 type of resistance to DDT. Resistance to dieldrin is apparently uniquely similar in all insect species (Oppenoorth 1965) and does not involve detoxication (Perry et al. 1964, Gerolt 1965, and others). Resistance to DDT conferred by the recessive gene kdr-O (for *knockdown resistance-Orlando*) and by similar genes (Milani 1960) in the house fly, *Musca domestica* L., is apparently independent of detoxication (Tsukamoto et al. 1965) as, most probably, are the incomplete recessive types of resistance to DDT in many species of mosquitoes (Georghiou 1965).

Although chlorinated insecticides are apparently toxic because of their effects on nervous systems, there is evidence that they may also interfere with various metabolic processes in animals. Thus, Tinsley (1964)

[1] Diptera: Muscidae.

[2] Approved as Technical Article 8422 by the Director, Texas Agricultural Experiment Station and supported in part by USDA Regional Research Project S-73. A portion of the work was done while the author was employed by the USDA, Corvallis, Oreg.

[3] The following abbreviations are used: G-6-P and G-6-PD for glucose-6-phosphate and G-6-P dehydrogenase, respectively: 6-PG and 6-PGD for 6-phosphogluconate and 6-PG dehydrogenase, respectively.

reported that glucose-6-phosphate dehydrogenase (G-6-PD) activity in rats was decreased when DDT was included in the diet. Tinsley (1966) demonstrated that treatment with dieldrin accentuated a fatty acid deficiency in the rat. Later, Tinsley (1968) indicated that dieldrin and thiamine exhibited antagonistic effects in rats.

In this report evidence is presented for what may prove to be significant variations in glucose metabolism which characterize certain house fly strains resistant to chlorinated hydrocarbon and cyclodiene insecticides. The investigation deals with measurements of 2 enzymes of the pentose phosphate cycle, glucose-6-phosphate dehydrogenase and 6-phosphogluconate dehydrogenase (6-PGD), enzymes previously studied in detail in the house fly by Chefurka (1957). Additional work involves measurements on the rate of production of $^{14}CO_2$ from differentially labeled glucose by flies of several strains both untreated and after treatment with the test insecticides.

MATERIALS AND METHODS.—*Insects.*—Strains of house flies used are listed below. Chromosomes are numbered according to Wagoner (1967). Mutant names are followed by appropriate symbols. Additional material on many of the strains has been published recently (Hoyer and Plapp 1966, 1968; Plapp and Hoyer 1967, 1968a, 1968b).

Orlando Regular—Susceptible strain used in many experiments.

SRS—World Health Organization standard reference strain. Susceptible.

Dieldrin-R — A phenotypically wild-type strain homozygous for a 4th chromosomal gene conferring resistance to dieldrin and other cyclodiene insecticides. Also resistant to DDT.

Dieldrin-R;*white⁵;stubby wing* (Dld-R;w⁵;stw/4:3:2) —Derived from the wild-type Dieldrin-R strain. Homozygous for resistance to dieldrin, but susceptible to DDT and other types of insecticides.

Dieldrin-R:*curly wing* (Dld-R:cyw/4:4) —Prepared by introducing dieldrin resistance from Dld-R;w⁵;stw strain into susceptible strain containing the 4th chromosomal recessive mutant *curly wing*.

Orlando DDT strain—Highly resistant to DDT. Homozygous for 2nd chromosome gene for DDT dehydrochlorinase and 3rd chromosomal gene kdr-O. Also homozygous for resistance to dieldrin.

DDT-R;*dot-vein* (DDT-R;dov/2:3) — Homozygous for a DDT dehydrochlorinase gene. Derived from Orlando DDT strain.

Knockdown Resistance-Orlando:white⁵ (kdr-O:w⁵/3:3) —Homozygous for a 3rd chromosomal gene conferring resistance to DDT and pyrethrins (Plapp and Hoyer 1968a). Derived from previously described kdr-O;stw strain (Hoyer and Plapp 1966) which was in turn derived from the Orlando DDT strain.

F_c—Obtained from F. J. Oppenoorth, Wageningen, The Netherlands. Homozygous for 5th chromosome gene DDT$_{md}$ (for *DDT microsomal detoxication*) conferring resistance to DDT and organophosphates (Oppenoorth 1967, Oppenoorth and Houx 1968).

Organotin-R;stubby wing (tin;stw/3;2) — Homozygous for a gene conferring resistance to organotin insecticides (Hoyer and Plapp 1968). Mechanism of resistance probably caused by decreased rate of absorption of insecticides (Plapp and Hoyer 1968b).

Chemicals. — Nicotinamide adenine dinucleotide (NADP), D-glucose-6-phosphate (disodium salt), and 6-phosphogluconic acid (barium salt) were purchased from Sigma Chemical Co., St. Louis, Mo. D-glucose-1-C^{14} and D-glucose-6-C^{14} were purchased in 50 μCi batches from Calbiochem, Los Angeles, Calif., 99% *p.p'* DDT was obtained from Geigy Chemical Co., Ardsley, N.Y., and 90% dieldrin from Shell Chemical Co., Modesto, Calif.

Test for Dehydrogenase Activity. — House fly homogenates were prepared by grinding individual female flies in ice-cold phosphate buffer (0.2 M, pH 7.4) in a Potter-Elvehjem type homogenizer. Female adults, 4–7 days old, were used at a concentration of 2 mg fresh weight/ml buffer. Homogenates were filtered through glass wool and usually assayed within 15 min of preparation.

G-6-P and 6-PG dehydrogenases were measured in the Beckman spectrophotometer, model DB, at room temperature. The reaction mixture consisted of 3 ml of fly homogenate (6 mg) and 0.1 ml of 0.2 m G-6-P or 6-PG. 0.1 ml of 0.02 m NADP was added to the sample immediately before the measurements of enzyme activity were started and the increase in absorption at 340 mμ caused by reduction of cofactor was measured at 30-sec or 1-min intervals. Under the conditions employed, the reactions were linear for at least 10 min. Results presented are based on readings obtained 1–6 min after addition of cofactor. In preliminary tests, the double substrate method of Glock and McLean (1955) was used to measure G-6-PD. However, no advantage occurred over measuring activity with G-6-P only, and their method was not used in subsequent tests.

Radiorespirometric Studies. — In radiorespirometric studies, the strains of flies tested were Orlando Regular (susceptible), kdr-O:w^5 (resistant to DDT and susceptible to dieldrin), and Dld-R:cyw (susceptible to DDT but resistant to dieldrin). Groups of 5 ♀ flies were anesthetized and injected intrathoracially with 2 μg ^{14}C-labeled glucose/fly (10^5 cpm). When insecticidal treatment was administered, the flies were treated topically on the tip of the abdomen with 1 μg DDT or dieldrin in 1 μliter acetone 30 min before injection of glucose. The dose of insecticide was not toxic to resistant flies but caused complete knockdown of susceptible flies, usually within 2 hr. All tests were repeated at least once, and several were repeated 3 or more times.

Immediately after the injection of glucose, flies

were placed in "respiration chambers." These chambers were 16×150-mm test tubes containing 1 ml 5% KOH. The flies were placed on a screen baffle about 12 cm above the KOH, and the tubes were plugged with cotton. At hourly intervals, the flies were transferred to clean tubes. Respired $^{14}CO_2$ trapped in the KOH was precipitated as $Ba^{14}CO_3$ and counted in a thin-window proportional counter.

RESULTS. — *Glucose-6-Phosphate Dehydrogenase.* — Measurements of G-6-PD were made over 4 months and thus represent averages obtained from several populations of each strain. The data (Table 1) show an interstrain variation of nearly 3-fold from high to low level of enzyme. Activity was highest in the strains homozygous for dieldrin resistance and in strains possessing the DDT resistance gene kdr-O. Activity was much lower in all other strains. This

Table 1.—Levels of glucose-6-phosphate dehydrogenase in several strains of house flies.

Strain tested	Resistance genes in test strains	No. of tests	Enzyme activity Change in O.D./min 6 mg fly ± SD
Dld-R;w⁵;stw	Dld-R	5	58±9
Dld-R:cyw	Dld-R	4	50±6
Dieldrin-R	Dld-R, Deh, kdr-O	3	49±4
kdr-O:w⁵	kdr-O	8	49±6
Orlando DDT	Deh, kdr-O, Dld-R	4	45±7
tin:stw	tin	4	29±4
Orlando Regular	—	10	28±7
SRS	—	3	27±5
F₀	DDT$_{md}$	4	24±9
DDT-R;dov	Deh	3	20±3

group included susceptible strains and strains (F₀, DDT-R;dov, and tin;stw) possessing other resistance genes.

Results obtained with the several strains resistant to dieldrin suggest very strongly that a positive relationship exists between resistance and high levels of G-6-PD. Enzyme activity was high in both the phenotypically wild-type strains resistant to dieldrin (Dieldrin-R and Orlando DDT) and in the 2 strains that combined resistance to dieldrin and visible mutants. These last 2 strains, Dld-R;w⁵;stw and Dld-R: cyw were derived sequentially from the Dieldrin-R strain. Dld-R;w⁵;stw was synthesized to obtain a strain resistant to dieldrin and susceptible to DDT (Plapp and Hoyer 1967). Dld-R:cyw was synthesized from a cross made between Dld-R;w^5;stw and a susceptible cyw strain to demonstrate linkage of Dld-R and cyw. Thus, the high G-6-PD level accompanied the gene for resistance to dieldrin through 2 transfers between strains. The most logical basis for this condition is that the same gene controls both resistance to dieldrin and high G-6-PD activity.

Similar results were obtained with strains that contained the DDT resistance gene kdr-O. Activity of G-6-PD was high, not only in the Orlando DDT strain but also in the kdr-O:w^5 strain. The latter strain was derived from a now-discarded kdr-O;stw strain (Hoyer and Plapp 1966) which was in turn derived from the Orlando DDT strain. Thus, kdr-O:w^5 bears the same relationship to the Orlando DDT strain that Dld-R: cyw bears to the Dieldrin-R strain. As with dieldrin resistance, DDT resistance associated with kdr-O and high G-6-PD levels appear to be genetically inseparable.

6-Phosphogluconate Dehydrogenase.—Measurements of 6-PGD are presented in Table 2 for 3 strains,

Table 2.—Levels of 6-Phosphogluconate dehydrogenase in 3 strains of house flies.

Strain tested	No. of tests	Enzyme activity Change in O.D./min/ 6 mg fly ± sd
Dld-R:cyw	4	35±7
kdr-O:w^5	4	30±10
Orlando Regular	4	26±7

Orlando Regular, kdr-O:w^5 and Dld-R:cyw, although preliminary measurements were made with all other test strains. With this enzyme, no significant difference in activity was found which characterized any particular type of resistant strain. Nevertheless, the relative rank of 6-PGD in the 3 strains is the same as in experiments with G-6-PD, that is, Dld-R:cyw > kdr-O:w^5 > Orlando Regular.

Respiration of C^{14}-Labeled Glucose.—Fig. 1 shows the radiorespirometric patterns for glucose-1-C^{14} and glucose-6-C^{14}. Results for untreated (no insecticide) susceptible flies are similar to those reported by Agosin et al. (1966) in that the rate of production of C^{14}O$_2$ was roughly identical after treatment with either C-1 or C-6-labeled glucose, a finding which indicates primary utilization of the glycolytic pathway for glucose metabolism. This is considered to be the case when there is no significant difference in respiration from the 2 glucose substrates.

The 2 resistant strains differed sharply from the susceptible strain in the patterns of glucose metabolism. Both kdr-O:w^5 and Dld-R:cyw initially oxi-

107

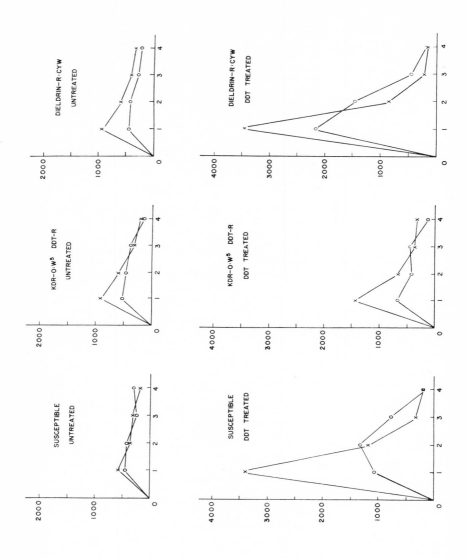

RECOVERED PER FLY

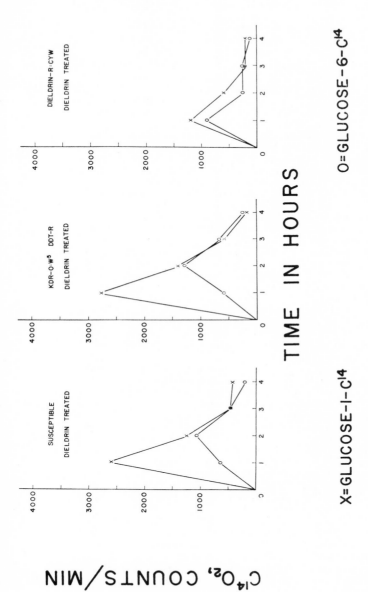

Fig. 1.—Patterns of $^{14}CO_2$ production in 3 strains of house flies untreated or after treatment with DDT or dieldrin.

X=GLUCOSE-1-C^{14} O=GLUCOSE-6-C^{14}

dized glucose-1-C^{14} at nearly twice the rate they oxidized glucose-6-C^{14}. Thus, these strains made relatively greater use of the pentose pathway. These results agree with the demonstrated findings of high levels of G-6-PD in these strains.

Large increases in respiration occurred in all strains after insecticidal treatment. Increases in $^{14}CO_2$ production were greatest after treatment with toxic doses of insecticide, e.g. dieldrin to kdr-O:w^5 flies, DDT to Dld-R:cyw flies, and both insecticides to Orlando Regular flies. Much smaller increases in rate of respiration occurred when flies were treated with insecticides to which they were resistant, nevertheless, some increase was always noted.

The stimulation of $^{14}CO_2$ production following insecticide treatment differed for the 2 labeled glucoses. Thus, the increase in $^{14}CO_2$ production from glucose-1-^{14}C was relatively greater (3–5 fold) than the increase in $^{14}CO_2$ production from glucose-6-^{14}C (2–4 fold). The net result of this differential stimulation is that while total respiration is always increased as a result of insecticidal treatment, the increase is relatively greater for the pentose pathway than it is for glycolysis.

In the case of DDT, these results confirm the findings of Silva et al. (1959) and Agosin et al. (1966) in that DDT treatment caused an increase in the relative rate of utilization of the pentose phosphate cycle. Effects of this type have not been reported previously with dieldrin. The dieldrin results were similar to those produced by DDT except that the increase in $^{14}CO_2$ from glucose-1-^{14}C was less sharp and the increase from Glucose-6-^{14}C was delayed as compared with DDT.

DISCUSSION. — Possible interrelationships between chlorinated insecticides and carbohydrate and oxidative metabolism in insects have been investigated in the past, often with contradictory and confusing results. A possible reason for this difficulty has been the less than adequate insect material available prior to the time that the advent of mutant house fly stocks made it possible to prepare genetically pure populations. Thus, to work with the DDT resistance gene kdr-O· it was necessary to prepare fly strains free of other DDT resistance genes. Similarly, investigations of dieldrin resistance have been facilitated by use of stocks free from DDT resistance.

The fact that high levels of G-6-PD are found in both kdr-O and Dieldrin-R strains does not imply a cause-effect relationship between enzyme level and resistance. This observation is based in part on the fact that kdr-O-type resistance to DDT, conferred by a 3rd chromosome gene, does not extend to dieldrin, nor does dieldrin resistance, conferred by a 4th chromosome gene, extend to DDT. Rather, it seems likely that whatever factor(s) result in resistance to DDT or dieldrin also result in an elevated G-6-PD level and a consequent high degree of utilization of

110

the pentose pathway in glucose metabolism.

Variations in G-6-PD levels occur readily in animals for a variety of reasons. Examples are the depression of G-6-PD activity in alloxan diabetic rats described by Glock and McLean (1955) as well as the previously mentioned decrease in G-6-PD activity in rats fed DDT (Tinsley 1964). Contrarily, starvation of rats followed by feeding of high carbohydrate diets results in G-6-PD levels of several times normal (Tepperman and Tepperman 1956), and stimulation of G-6-PD activity has been reported as a consequence of ethionine treatment of rats (Sie and Fishman 1968). The significance of these changes, both in terms of chlorinated insecticides and in terms of other types of stress are not well understood.

REFERENCES CITED

Agosin, M., B. C. Fine, N. Scaramelli, J. Ilivicky, and L. Aravena. 1966. The effect of DDT on the incorporation of glucose and glycine into various intermediates in DDT-resistant strains of *Musca domestica* L. Comp. Biochem. Physiol. 19: 339–49.

Chefurka, W. 1957. Oxidative metabolism of carbohydrates in insects. II. Glucose-6-phosphate dehydrogenase and 6-phosphogluconate dehydrogenase in the house fly *Musca domestica* L. Enzymologia 17: 209–26.

Georghiou, G. P. 1965. Genetic studies on insecticide resistance. Advan. Pest Contr. Res. 10: 171–230.

Gerolt, P. 1965. The fate of dieldrin in insects. J. Econ. Entomol. 58: 849–57.

Glock, G. E., and P. McLean. 1955. A preliminary investigation of the hormonal control of the hexose monophosphate oxidative pathway. Biochem. J. 61: 390–7.

Hoyer, R. F., and F. W. Plapp, Jr. 1966. A gross genetic analysis of two DDT-resistant house fly strains. J. Econ. Entomol. 59: 495–501.
——. 1968. Insecticide resistance in the house fly: Identification of a gene that confers resistance to organotin insecticides and acts as an intensifier of parathion resistance. Ibid. 61: 1269–76.

Milani, R. 1960. Genetic studies on insecticide-resistant insects. Misc. Publ. Entomol. Soc. Amer. 2: 75–83.

Oppenoorth, F. J. 1965. Biochemical genetics of insecticide resistance. Annu. Rev. Entomol. 10: 185–206.
——. 1967. Two types of sesemex-suppressible resistance in the house fly. Entomol. Exp. Appl. 10: 75–86.

Oppenoorth, F. J., and N. W. H. Houx. 1968. DDT resistance in the house fly caused by microsomal degradation. Ibid. 11: 81–93.

Perry, A. S., G. W. Pearce, and A. J. Bucher. 1964. The absorption, distribution and fate of C^{14}-aldrin and C^{14}-dieldrin by susceptible and resistant house flies. J. Econ. Entomol. 57: 867–72.

Plapp, F. W. Jr., and R. F. Hoyer. 1967. Insecticide resistance in the house fly: Resistance spectra and preliminary genetics of resistance in eight strains. Ibid. 60: 768–74.
——. 1968a. Possible pleiotropism of a gene conferring resistance to DDT, DDT analogs, and pyrethrins in the house fly and *Culex tarsalis*. Ibid. 61: 761–5.

111

1968b. Insecticide resistance in the house fly: Decreased rate of absorption as the mechanism of action of a gene that acts as an intensifier of resistance. Ibid. 61: 1298–1303.

Sie, H., and W. H. Fishman. 1968. Stimulation of pentose phosphate pathway dehydrogenase enzyme in ethionine-treated mice. Biochem. J. 106: 769–76.

Silva, G. M., W. P. Doyle, C. H. Wang. 1959. Glucose catabolism in the DDT treated American cockroach. Arq. Port. Biochim. 1959: 1–7.

Tepperman, H. M., and J. Tepperman. 1956. The hexosemonophosphate shunt and adaptive hyperlipogenesis. Diabetes 7: 478–85.

Tinsley, I. J. 1964. Ingestion of DDT and liver Glucose-6-phosphate dehydrogenase activity. Nature (London) 202: 1113–4.
    1966. Nutritional interactions in dieldrin toxicity. J. Agr. Food Chem. 14: 563–5.
    1968. An interaction of dieldrin with thiamine. Proc. Soc. Exp. Biol. Med. 129: 463–5.

Tsukamoto, M., T. Narahashi, and T. Yamasaki. 1965. Genetic control of low nerve sensitivity in insecticide-resistant house flies. Botyu-Kagaku 30: 128–32.

Wagoner, D. E. 1967. Linkage group-karyotype correlation in the house fly determined by cytological analysis of X-ray induced translocations. Genetics 57: 729–39.

112

Susceptibility of Adult *Hippelates pusio*[1] to Insecticidal Fogs[2]

R. C. AXTELL and T. D. EDWARDS

Hippelates eye gnats are readily attracted to man and animals, can be extremely annoying, and are possible vectors of eye diseases. Control of these gnats is desired in resort areas, especially around golf courses, where their annoyance of people creates a distinct economic problem. Repellents are useful for temporary relief (Axtell 1967) and it may be possible to utilize residual insecticide sprays for gnat control (Axtell and Edwards 1970). Thermal fogs of DDT, lindane, and malathion while being used in mosquito control operations have been observed to simultaneously give poor control of eye gnats (Jones and Magy 1951, Rees 1952, Mulla 1959). Standardized tests comparing several insecticidal fogs against eye gnats have not been reported. Therefore, several insecticides were evaluated as thermal fogs against adult *H. pusio* Loew, the most common pestiferous species of eye gnat in the southeastern United States.

MATERALS AND METHODS.—The *H. pusio* used were from colonies (maintained at 26.7°C and 55% RH) established from several hundred adults collected in North Carolina in 1963–64. The larvae were reared in a mixture of CSMA (Chemical Specialties Manufacturers Association) medium[3] and vermiculite. The adults' food consisted of dried beef blood, honey, and strained prunes. Rearing and handling details have

---

[1] Diptera: Chloropidae.
[2] Contribution from the Entomology Department, North Carolina State University Agricultural Experiment Station, Raleigh.

[3] Ralston Purina Co., St. Louis, Mo.

been described (Axtell 1964). They were modifications of the techniques used by Bay and Legner (1964). Adult gnats were exposed to the insecticidal fog 3–5 days after emergence from the puparia.

The following testing procedure was used. After anesthetization with CO_2, 150 gnats were placed in each exposure cage in the laboratory. These cages were round (13 cm diam $\times$ 1 cm thick) metal embroidery hoops with 32-mesh screen. The bottom screen was fastened in place with epoxy glue on the inner hoop and the top screen was held in place by the tight-fitting outer hoop. The caged gnats were suspended outdoors at 1 m above the ground in a vertical orientation (the metal rim of the cage perpendicular to the ground) at a distance of 8 m from the fogging machine. The proper orientation of the cages to the machine and wind direction was determined immediately prior to testing so that the fog would pass through the cages uniformly. Cages were exposed to the fog for 1 min. Then the gnats were immediately anesthetized with CO_2 at the treatment site and placed in glass holding jars (0.5 liter) capped with 40-mesh screen tops. To provide food, a cotton pad soaked in 10% sucrose solution was placed on the screen top. The holding jars were returned to the laboratory and held in an incubator at 21°C. (Travel time between the laboratory and the field testing site was about 10 min.) Dead gnats (those unable to walk) were counted at 2 and 24 hr after treatment. Untreated gnats were handled in the same manner, including being transported to and from the field testing site, but they were not hung in the insecticidal fog.

Each insecticide was evaluated in 4–9 tests on different days with the air temperature 23°–31°C. Each test included 4 treated cages of gnats for each of 5 concentrations of insecticide and 4 untreated cages. Mortalities in the untreated controls was routinely less than 2% and 4% at 2 and 24 hr posttreatment, respectively. Therefore, mortalities among the treated gnats were not corrected for control mortalities.

The insecticidal fog was produced by a portable Swingfog S. N. 7 machine with an 0.8-mm-diam flow-control jet giving an output of 125 ml/min. The insecticides were diluted in no. 2 fuel oil to the desired concentration of active ingredient (% on a wt/wt basis). In separate tests, no gnat mortality resulted from exposure to fog containing the fuel oil alone. The identities, formulations, and sources of insecticides are presented at the end of this report.

RESULTS AND DISCUSSION.—Table 1 presents gnat mortalities from exposure to the various concentrations of insecticides. In general, the 24-hr mortalities were only slightly greater than the 2-hr values. DDT, the least effective insecticide, exhibited the greatest difference between 2 and 24-hr posttreatment mortalities. The most effective insecticides were naled and propoxur with 2-hr-posttreatment mortalities of 92% and 94%, respectively, at 0.1% concentration.

Table 1.—Mortalities of *H. pusio* exposed to insecticidal fogs for 1 min.

Chemical	No. tests	Mean % mortality at various concentrations (%)						Estimated LC₅₀ (%)
		1.0	0.5	0.1	0.01	0.001	0.0	LC_{50} (%)
		2 hr after exposure						
DDT	9	28.7	38.9	17.2	0.7	1.4	1.6	>1.0
Fenthion	4	98.3	100	6.8	1.4	1.1	1.8	0.19
Gardona	4	100	98.8	18.4	5.6	0.5	0.4	.16
Dursban	4	99.1	100	15.7	.7	1.9	1.3	.16
Dichlorvos	5	100	100	24.6	8.4	3.1	1.3	.14
C-9491	7	100	100	39.6	10.4	8.8	.8	.12
Malathion + Lethane[a]	4	97.7	99.3	37.4	2.1	7.3	1.2	.12
Propoxur	5	100	100	94.5	2.0	3.4	1.2	.035
Naled	5	100	100	92.2	14.3	.9	.9	.027
		24 hr after exposure						
DDT	9	65.1	52.5	23.5	13.5	18.3	3.7	.5
Fenthion	4	100	100	28.1	2.7	2.3	2.8	.14
Gardona	4	100	99.6	35.8	20.8	7.8	2.8	.12
Dursban	4	100	100	27.8	2.1	3.1	2.3	.14
Dichlorvos	5	100	100	32.3	16.9	4.1	3.7	.13
C-9491	7	100	100	44.2	27.8	23.8	3.0	.11
Malathion + Lethane[a]	4	99.8	99.7	47.0	5.7	11.2	2.1	.11
Propoxur	5	100	100	91.1	3.3	4.7	2.5	.035
Naled	5	100	100	99.6	18.4	1.1	2.6	.02

[a] Concentrations refer only to % malathion. Dilutions were made from a formulation containing 44% malathion and 20% Lethane 384.

Dichlorvos, C-9491, and the malathion + Lethane 384 mixture were considerably less effective, based on the 2-hr-posttreatment mortalities, while fenthion, Dursban, and Gardona were even less effective. These differences declined in the 24-hr-posttreatment mortalities. Mortality counts after 24 hr showed C-9491 and the mixture of malathion and Lethane 384 to be slightly more effective than fenthion, dichlorvos, Dursban, and Gardona, which were about equal to one another in effectiveness. To facilitate comparisons of the effectiveness of these insecticides, the LC$_{50}$ values were extrapolated (by log-probability graphing) and are presented in Table 1. These values are crude estimations because of the limited number of treatments producing mortalities in the middle range of the dosage-response curve.

Fogging has been used rountinely for control of biting Diptera, especially mosquitoes. The relative effectiveness of insecticidal fogs against eye gnats may be different than for other Diptera. DDT has often been used in fogs for mosquito control but was quite ineffective against the eye gnats. Naled and propoxur were the most effective insecticidal fogs for gnat control. However, the other insecticides tested were effective enough to warrant further evaluation. Large-scale fogging tests should be conducted against field populations of eye gnats. Nonthermal fogs and mists should also be evaluated.

IDENTITIES, FORMULATIONS, AND SOURCES OF INSECTICIDES.—
DDT, 30% (w/w) T in napthalene, Grace Chemical Co.
Naled, 14 lb/gal solution, Chevron Chemical Co.
Fenthion, 8 lb/gal oil conc, Chemagro Corp.
Propoxur, 10% oil conc, Chemagro Corp.
Dichlorvos, 1% oil conc, Shell Chemical Co.
Dursban, O,O-diethyl O-3,5,6-trichloro-2-pyridyl phosphorothioate, 6 lb/gal oil conc, Dow Chemical Co.
Gardona®, 2-chloro-1- (2,4,5-trichlorophenyl) vinyl dimethyl phosphate, 2 lb/gal EC, Shell Chemical Co.
CIBA C-9491, O- (2,5-dichloro-4-iodophenyl) O,O-dimethyl phosphorothioate, 2 lb/gal oil conc, CIBA Corp.
Malathion (44%) + Lethane 384 (20%) (2- (2-butoxyethoxy) ethyl thiocyanate), oil solution, Forshaw Chemical Co., Charlotte, N. C.

REFERENCES CITED

Axtell, R. C. 1964. Laboratory rearing of *Hippelates pusio* and *H. bishoppi* (Diptera:Chloropidae). J. Elisha Mitchell Sci. Soc. 80: 163–4.
1967. Evaluations of repellents for *Hippelates* eye gnats. J. Econ. Entomol. 60 (1) : 176–80.
Axtell, R. C., and T. D. Edwards. 1970. Susceptibility

of adult *Hippelates* eye gnats to insecticidal deposits. Ibid. 63:

Bay, E. C., and E. F. Legner. 1964. Quality control in the production of *Hippelates collusor* (Tsnd.) for use in the search and rearing of their natural enemies. Proc. Amer. Mosquito Control Ass. 19: 403–10.

Jones, R. W., and H. I. Magy. 1951. Mosquito control techniques for control of *Hippelates* spp. (Diptera: Chloropidae). Mosquito News 11: 102–7.

Mulla, M. S. 1959. Some important aspects of *Hippelates* gnats with a brief presentation of current research findings. Proc. Calif. Mosquito Control Ass. 27: 48–52.

Rees, B. E. 1952. The *Hippelates* fly or eye gnat in Fresno, California, and the feasibility of its control by present mosquito abatement methods. Ibid. 20: 85–87.

DDT in Birds

Influence of Iodinated Casein on DDT Residues in Chicks[1]

W. E. DONALDSON, M. D. JACKSON AND T. J. SHEETS

Department of Poultry Science & Pesticide Residue Research Laboratory,
North Carolina State University, Raleigh, North Carolina 27607

OLNEY *et al.* (1962) showed that chlorinated hydrocarbon pesticides tend to accumulate in the fatty deposits of chickens. Wesley *et al.* (1966) presented evidence which suggested that starvation increased the depletion rate of DDT from the tissues of chickens. Repeated cycles of 48 hours of starvation followed by 24 hours of feeding non-contaminated feed reduced body fat content and body burden of DDT when starved chicks were grown to the same body weight as full-fed controls (Donaldson *et al.*, 1968). The results with starvation suggest that any treatments which tend to decrease body fat may also tend to increase the rate of elimination of DDT residues from the body. Turner *et al.* (1946) showed that feeding iodinated casein to laying hens resulted in decreased deposition of subcutaneous and carcass fat. Stamler *et al.* (1950) found that the feeding of dessicated thyroid to chicks reduced the lipemia and fatty liver produced by cholesterol feeding. The experiments to be reported here were designed to measure the influence of iodinated casein on DDT residues in chicks when dilution effects attributable to body size were eliminated as a variable.

MATERIALS AND METHODS

Two experiments were conducted in which female White Rock × Cornish chicks were fed a diet containing 100 p.p.m. DDT for 2 weeks. Feed and water were available *ad libitum* throughout both

experiments except during starvation when only water was available. The feed was a commercial starter diet obtained locally. Representative samples of the feed were analyzed for DDT. No significant levels of DDT were found in any feed used subsequently to DDT dosing of the chicks. Prior to initiation of the experiments, p,p'-DDT (99 + %) dissolved in corn oil (5 gm./ ml.) was added to the feed.

In experiment 1, one-day old chicks were divided randomly into 16 groups of 10 chicks each. All groups received feed containing 100 p.p.m. of DDT for 2 weeks. After 2 weeks, the groups received DDT-free feed until killed. Eight of the 16 groups were fed the control feed with no additions, and the remaining 8 groups were fed the control feed with 0.03% iodinated casein (Nutritional Biochemicals, Cleveland, Ohio). Four of the groups receiving each diet were full-fed for 2 weeks and killed. The remaining 4 groups receiving each diet were fed in 3-day cycles of 2 days starvation and 1 day full-fed until their average body weights approximated the final weights of the full-fed groups receiving the corresponding diets. These groups were then killed. A 2-gm. sample of abdominal adipose tissue was removed from each bird. The samples were pooled for each 10-bird group and were frozen until analyzed. The feathers were removed from each bird, and the carcass was blended with distilled water in a Waring blendor. Duplicate 10-gm. samples were removed and pooled such that there were 2 samples of 100 gm. each per group of 10 birds. These were frozen until analyzed for moisture and fat content.

[1] Paper no. 3380 of the Journal Series of the North Carolina State University Agricultural Experiment Station, Raleigh, North Carolina.

TABLE 1.—*Effects of periodic starvation* vs. *continuous feeding of diets with and without iodinated casein on body weight, body fat and DDT residues (Experiment 1)*

Treatment[1] Dietary iodinated casein, %	Feeding system	Age, days	Body wt., grams[2]	Body fat, %[3]	Adipose tissue levels, p.p.m.[3]				Estimated body burden, mg./bird[4]			
					DDE	DDD	DDT	Total	DDE	DDD	DDT	Total
0	Full-fed	28	403	9.1	107	46	204	357	3.9	1.6	7.4	12.9
0.03	Full-fed	28	400	6.5	142	40	263	445	3.7	1.0	6.8	11.5
0	Starved	38	404	6.9	206	27	145	378	5.7	0.7	4.1	10.5
0.03	Starved	41	419	6.3	307	18	131	456	8.1	0.4	3.4	11.9

[1] All chicks were fed a diet containing 100 p.p.m. of DDT for the first 14 days.
[2] Each value based on 40 chicks.
[3] Each value based on pooled samples from 4 groups of 10 chicks each.
[4] These estimations are based on the assumption that the residues were similarly distributed among all body fat pools.

The procedure for experiment 2 was identical to that for experiment 1 except that the post-DDT treatment period for the full-fed groups was 3 weeks and dietary levels of iodinated casein were 0, 0.03 and 0.06%.

p,p'-DDT and its metabolites, p,p'-DDE and p,p'-DDD were determined in adipose tissue and feed as described previously (Donaldson *et al.*, 1968). Known amounts of the pesticides were added to samples which were analyzed for recovery determinations. Average recoveries of 92, 100 and 106%, for DDE, DDD and DDT, respectively, were obtained.

Carcass samples, which were used for moisture and fat determinations, were dried to a constant weight at 80°. The dried samples were extracted with diethyl ether for 3 hours in a Goldfisch extractor. The solvent was evaporated at 60° under a stream of nitrogen and the extracts were weighed.

All statistical analyses were based upon analysis of variance for factorial experiments. These methods were outlined in Snedecor (1948).

RESULTS

The results of experiment 1 are shown in Table 1. Both periodic starvation and

0.03% of dietary iodinated casein (I.C.) reduced body fat compared with full-fed controls, although the average body weights of all treatments were similar. The concentration (p.p.m. in adipose tissue) of DDE was significantly increased ($P < 0.01$) by starvation and I.C. The interaction between these factors was also significant ($P < 0.01$). Tissue concentration of DDD and DDT were decreased by starvation ($P < 0.01$) but not by I.C. ($P > 0.05$). I.C. significantly increased tissue concentrations of total residues ($P > 0.01$) but starvation was without effect ($P > 0.05$). The body burden (mg. of residue/bird) of DDE was not affected by I.C. in full-fed birds ($P > 0.05$), but was significantly increased by I.C. in starved birds ($P > 0.01$). Both starvation and I.C. reduced body burdens of DDD ($P < 0.01$). Starvation significantly reduced burdens of DDT ($P < 0.01$), but the small reductions observed with I.C. in fed and starved birds were not statistically significant ($P > 0.05$).

Although I.C. had no statistically significant effects on burdens of DDT, there were small reductions, which might be magnified with increased time and/or level of feeding of I.C. Therefore, in experiment 2, length of time of full-feeding subsequent to DDT

121

TABLE 2.—*Effects of periodic starvation* vs. *continuous feeding of diets with varying levels of iodinated casein on body weight, body fat and DDT residues (Experiment 2)*

Treat-ment[1] Dietary iodinated casein, %	Feeding system	Age, days	Body wt., grams[2]	Body fat, %[3]	Adipose tissue levels, p.p.m.[3]				Estimated body burden, mg./bird[4]			
					DDE	DDD	DDT	Total	DDE	DDD	DDT	Total
0	Full-fed	35	566	14.0	46	30	110	186	3.6	2.4	8.7	14.7
0.03	Full-fed	35	518	12.7	76	45	119	240	5.0	3.0	7.8	15.8
0.06	Full-fed	35	527	9.8	60	39	120	219	3.1	2.0	6.2	11.3
0	Starved	54	555	9.2	98	23	101	222	5.0	1.2	5.2	11.4
0.03	Starved	57	537	4.1	179	31	42	252	4.0	0.7	0.9	5.6
0.06	Starved	54	525	5.7	103	10	16	129	3.1	0.3	0.5	3.9

[1] All chicks were fed a diet containing 100 p.p.m. of DDT for the first 14 days.

[2] Each value based on 40 chicks.

[3] Each value based on pooled samples from 4 groups of 10 chicks each.

[4] These estimations are based on the assumption that the residues were similarly distributed among all body fat pools.

dosing was increased to 3 weeks, and the additional I.C. level of 0.06% was included in the design. The results of experiment 2 are presented in Table 2.

As in experiment 1, both starvation and I.C. produced birds with lower body fat content. Starvation resulted in significantly higher (P < 0.01) concentrations of DDE and lower concentrations of DDD and DDT in adipose tissue when the periodically starved groups were compared with the full-fed groups fed the same I.C. level. These changes in concentrations offset one another in the control and 0.03% I.C. groups such that total residue concentrations were not affected by starvation (P > 0.05). When 0.06% I.C. was fed, the relative increase of DDE concentration induced by starvation was smaller and the relative decrease of DDT concentration induced by starvation was greater than the corresponding starvation induced changes in the groups fed the other diets. Hence, in the 0.06% I.C. groups, starvation significantly lowered total residue concentration (P < 0.01). I.C. caused a slight rise in DDE, DDD and total residue concentra-

tions in full-fed birds but did not affect DDT concentration. When I.C. was fed to starved animals, the results differed with dose levels of I.C. In starved groups fed I.C., a rise in DDE and a fall in DDT concentrations were observed in comparison to starved controls. These changes were offsetting. In starved groups fed 0.06% I.C., DDE concentration was comparable to the concentration in starved controls, but DDT concentration was drastically reduced. This interaction was highly significant (P < 0.01).

Table 2 also shows the data on body burden of residues. Starvation increased the body burden of DDE in birds fed the control diet but not in birds fed the I.C. diets (P < 0.05). Starvation decreased the body burdens of DDT in all cases, but the effects were most pronounced when the diets contained I.C. I.C. at both levels decreased the body burdens of DDT in both full-fed and starved (P < 0.01). The interaction was highly significant (P < 0.01). The total residue burdens tended to parallel those of DDT except in the groups full-fed the 0.03% I.C. diet.

122

DISCUSSION

These results indicate that I.C. is an effective dietary treatment for increasing the rate of elimination of DDT and associated residues from the body. The effect of I.C. was more pronounced in periodically starved than in full-fed birds. Starvation and 0.03% I.C. tended to increase body burdens of DDE. The increase with 0.03% I.C. was not as pronounced in experiment 2, and 0.06% I.C. did not elicit the increase. The DDE data suggest that both starvation and I.C. tend to speed the conversion of DDT to the less toxic DDE. Higher levels of I.C. feeding (0.06%) or feeding 0.03% I.C. for longer periods to starved birds (compare Table 1 with Table 2), resulted in the elimination of the DDE residue. The data presented here further suggest that the mechanism responsible for increased elimination rates of DDT is similar for both starvation and I.C. feeding (*i.e.* increased mobilization of body fat stores and the consequent greater availability of the DDT residues associated with body fat to degradative processes). This interpretation is supported by the work of Hunt (1966) who stated that pheasants could be maintained for extended periods on a diet containing 100 p.p.m. DDT with no ill effects, but that pheasants fed 10 p.p.m. DDT for the same period experienced mortality when feed intake was curtailed.

The results presented here re-emphasize the view presented earlier (Donaldson *et al.*, 1968) that accidental contamination of chickens with persistent pesticides, if recognized prior to marketing, need not cause condemnation because of excess pesticide residue. A feeding program employing I.C. and the judicious use of periodic starvation with subsequent full-feeding to market weight can result in tissue concentrations of residues that are considered acceptable. This view assumes that reasonable tolerances for such pesticides and their metabolites are in effect at time of marketing.

SUMMARY

Two experiments were conducted to compare the effects of diets containing 0, 0.03 or 0.06% iodinated casein (I.C.) on the elimination of DDT residues from full-fed and periodically starved chicks which had been dosed previously with DDT. Periodic starvation and dietary IC decreased body fat content and tended to increase the conversion of DDT to DDE. The changes in DDE and DDT concentrations in adipose tissue of starved and 0.03% I.C.-fed birds were offsetting, and hence total residue concentrations were not affected greatly by the treatments. When 0.03% I.C. was fed for longer periods or when 0.06% I.C. was fed, elimination of DDE appeared to be stimulated in starved birds; and concentrations of total residue were lower than in starved birds which did not receive I.C. Similar patterns were observed for estimated body burden of residues. Because of the lower body fat content of starved and I.C.-treated birds, body burdens of the residues in these groups were lower than in comparable control groups even when concentration values were equal to or higher than control values. The effects of I.C. were more pronounced in starved than in full-fed birds.

ACKNOWLEDGEMENT

The technical assistance of Mr. C. Strickland is gratefully acknowledged.

REFERENCES

Donaldson, W. E., T. J. Sheets and M. D. Jackson, 1968. Starvation effects on DDT residues in chick tissues. Poultry Sci. 47: 237–243.

Hunt, E. G., 1966. Biological magnification of pesticides. In Scientific Aspects of Pest Control, Publication No. 1402 NAS-NRC, p. 257.

Olney, C. E., W. E. Donaldson and T. W. Kerr, 1962. Methoxychlor in eggs and chicken tissues. J. Econ. Entomol. 55: 477–479.

Snedecor, G. W., 1948. Statistical Methods. Iowa State College Press, Ames, Iowa. 4th ed., pp. 400–408 and 423–430.

Stamler, J., C. Bolene, E. Levinson and M. Dudley, 1950. The lipotropic action of dessicated thyroid in the cholesterol-fed chick. Endocrinol. 46: 382–386.

Turner, C. W., H. L. Kempster and N. M. Hall, 1946. Effect of continued thyroprotein on egg production. Poultry Sci. 25: 562–569.

Wesley, R. L., A. R. Stemp, W. J. Stadelman, B. J. Liska, R. L. Adams and R. B. Harrington, 1966. Further studies on depletion of DDT residues from commercial layers. Poultry Sci. 45: 1136.

Comparison of DDT Effect on Pentobarbital Metabolism in Rats and Quail

Joel Bitman, Helene C. Cecil, Susan J. Harris, and George F. Fries

I n almost all species the liver is the major site for the metabolism of foreign chemicals (Bush, 1963). The barbiturates are a class of foreign chemicals which are strong depressants of the central nervous system and induce sleep in the treated individual. A standard dose of pentobarbital will produce a standard sleeping time, and the duration of this period is primarily dependent upon the detoxification of the barbiturate by the liver. An increase in those liver microsomal enzymes which metabolize the barbiturate will be reflected in a shorter sleeping time. Conversely, any inhibition of the activity of these liver microsomal enzymes should be reflected in a longer duration of barbiturate sleeping time.

125

Hart *et al.* (1963) accidentally discovered that spraying of animal rooms with organochlorine pesticides reduced sleeping time in the rat, suggesting that the liver microsomes had been stimulated. Subsequent studies by these investigators and others (Hart *et al.*, 1963; Hart and Fouts, 1965; Conney *et al.*, 1967) have demonstrated clearly that the microsomal enzyme systems exhibited increased metabolic activity *in vitro* for several foreign drugs and that *in vivo* the sleeping times were shortened with DDT, DDE, rhothane, perthane, methoxychlor, and chlordane.

It has recently been shown that the *o,p'*-isomer of DDT, which constitutes 15–20% of technical DDT, exerts an estrogen-like action on the reproductive tissues of mammals and birds (Levin *et al.*, 1968; Welch *et al.*, 1969; Bitman *et al.*, 1968). In contrast, *p,p'*-DDT has little if any estrogenic activity. Since *p,p'*-DDT shortens pentobarbital sleeping time, it seemed of interest to determine whether the *o,p'*-isomer induces a similar effect in rats. Accordingly, we studied in detail the time sequence of the effect upon pentobarbital sleeping time of the *o,p'*- and *p,p'*-isomers of DDT, and its two major metabolic products, DDE and DDD. The effects of treatment on mammals and birds were compared by conducting these experiments on rats and Japanese quail.

MATERIALS AND METHODS

Male and female rats (*Rattus norvegicus*) of an inbred Wistar strain (200–300 g of body weight) on a schedule of 12 hr of light and 12 hr of dark were fed diets containing the *o,p'*- and *p,p'*-isomers of DDT, DDE, and DDD (99+%, Aldrich Chemical Co.). Immature rats were placed on diets at 21 days of age (35–45 g body weight). Female rats were ovariectomized at 50 days of age; after a recovery period of 14 days they were placed on the appropriate pesticide-containing diet. Male rats were castrated when 36 days old; they were injected subcutaneously with 2 mg testosterone propionate/0.2 ml olive oil or 5 μg estradiol-17β three times a week for the duration of the experiment. The pesticides were fed after a hormone pretreatment period of 2 weeks.

Japanese quail (*Coturnix coturnix japonica*), on a schedule of 14 hr of light and 10 hr of dark, were fed *ad libitum* diets containing the pesticides. The quail were $1^1/_2$ to 3 months of age when experiments were initiated. A corn-soybean meal type starter ration was used as the basal diet. Pesticides were dissolved in acetone and pipetted onto the diet, which was spread out in shallow pans. After evaporation of the solvent, the diets were thoroughly mixed.

The sleeping time was defined as the time between injection of the barbiturate and restoration of the righting reflex.

Sodium pentobarbital was injected intraperitoneally in rats at a dosage rate of 32 mg/kg body weight. A dosage of 75 mg/kg of sodium pentobarbital was administered by the intramuscular route in Japanese quail.

Following florisil cleanup, concentrations of the pesticides in the lipid were determined by gas-liquid chromatography (glc). An F & M model 810 instrument equipped with an electron capture detector was used. The 4-ft $\times$ $^1/_4$ in. o.d. glass column was packed with 3.8% silicon gum rubber on 80-100 mesh Chromosorb W (acid-washed dimethylchloro-silane treated). A 95% argon–5% methane mixture was used as both carrier and purge gas. The column temperature was 185° C. Under these conditions the relative retention times were: o,p'-DDE, 1.00; o,p'-DDD, 1.30; o,p'-DDT, 1.75; p,p'-DDE, 1.26; p,p'-DDD, 1.66; and p,p'-DDT, 2.25. Thus, it was not possible to distinguish between o,p'-DDD

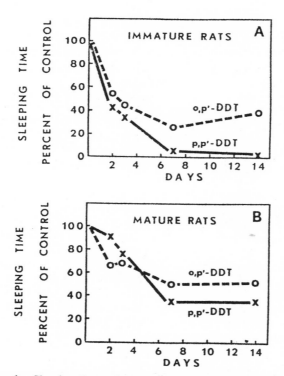

Figure 1. Sleeping times of rats fed 50 ppm o,p'- or p,p'-DDT. A. Immature male and female rats. Each point represents 8–12 rats. B. Intact cycling and ovariectomized female rats. Each point represents 8–12 rats

127

and p,p'-DDE, or p,p'-DDD and o,p'-DDT by glc.

Representative samples from each experimental group were checked by thin-layer chromatography (tlc) on aluminum oxide G plates in which silver nitrate was incorporated for visualization (A.O.A.C., 1966). Parallel standards of the various analogs were also spotted. The R_f's of the six analogs were: o,p'-DDE, 0.51; o,p'-DDD, 0.23; o,p'-DDT, 0.49; p,p'-DDE, 0.54; p,p'-DDD, 0.20; and p,p'-DDT, 0.39. The spots were scraped from the thin-layer plates and the pesticides recovered from the adsorbent by elution with petroleum

	Percent of Control Sleeping Time After Feeding for	
Group	7 Days	14 Days
Immature Females		
p,p'-DDT	10 ± 2	4 ± 1
o,p'-DDT	37 ± 3	39 ± 2
p,p'-DDE	1 ± 0.2	7 ± 2
o,p'-DDE	13 ± 6	14 ± 1
p,p'-DDD	28 ± 4	65 ± 7
o,p'-DDD	40 ± 9	44 ± 4
Mature Females		
p,p'-DDT	51 ± 5	41 ± 5
o,p'-DDT	63 ± 4	60 ± 5
p,p'-DDE	47 ± 5	24 ± 3
o,p'-DDE	31 ± 5	24 ± 3
p,p'-DDD	77 ± 1	57 ± 5
o,p'-DDD	68 ± 2	61 ± 7

Table I. DDT Analogs and Pentobarbital Sleeping Time in Rats

Immature Females: Each group consisted of five rats treated with 50 ppm of the pesticide.
Mature Females: Each group contained ten rats treated with 100 ppm of the pesticide.

ether. The identity of each pesticide was then reconfirmed by glc.

The treated birds were fed only a single compound. Tlc gave no evidence that both members of the pairs that were indistinguishable by glc were present within the same sample. Thus, no complications arose from using this glc method for the quantitative determinations.

RESULTS AND DISCUSSION

Two days after immature 21-day old rats were placed on a diet containing 50 ppm of the DDT isomer, the duration of pentobarbital anesthesia was reduced to only half that of controls (Figure 1A). There were no differences between immature male and female rats, and the line shown represents combined data from both sexes. After 1 week on the DDT diet, the p,p'-fed rats had sleeping times which were only 5% of controls, while the o,p'-DDT fed rats showed a pentobarbital anesthesia of about 30–40% of controls. In immature rats it appears that p,p'-DDT induces liver microsomal enzyme activity which almost completely metabolizes the standard dose of pentobarbital, while o,p'-DDT fed at the same level induces liver enzyme activity to a lesser extent, and sleeping time reductions of only 60–70% are observed.

Figure 1B shows the sleeping time reduction produced by o,p'- and p,p'-DDT in mature female rats, either intact cycling females or ovariectomized females. There were no differences between these two groups and their sleeping time data were combined. The 50 ppm p,p'-DDT diet has less effect in the mature than in the immature rats, reducing sleeping time to only 75% at 2–3 days and achieving a maximal reduction to about $1/3$ of controls. The o,p'-DDT fed rats showed a similar reduction to about 50% of control levels.

The effects upon pentobarbital sleeping time of the isomers of the DDT analogs, DDE and DDD, were also studied in immature and mature female rats. Table I shows the determination of sleeping time performed on rats after being on the diet for 1 and 2 weeks. With both the immature and mature rats, DDE appeared to elicit a greater reduction than DDT, which in turn was more effective than DDD. All compounds, however, stimulated pentobarbital metabolism and sleeping times were reduced to 5–60% of controls. It appears that the o,p'-isomer was about as effective as the p,p'-isomer for all compounds except DDT, in which case the p,p'-isomer lowered sleeping time to a much greater extent.

Although the immature male and female rats showed little or no differences in sleeping time, and the data shown in Figure

129

1A represented combined data, there were differences in male rats as they matured. Mature male rats showed much less response to 100 ppm o,p'-DDT in reduction of sleeping time than did the immature male rats fed 50 ppm o,p'-DDT (Figure 2). Conney and Burns (1962) have also found that foreign chemicals were able to stimulate drug-metabolizing enzymes to a greater extent in the immature male than in the adult male rat. If the mature rats were castrated and placed on a 100 ppm o,p'-DDT diet, they behaved somewhat like immature male rats, showing much greater reductions in pentobarbital sleeping time. When these castrates were treated with estrogen, they still acted like castrates as far as sleeping time was concerned, indicating that the female sex hormone does not affect the liver microsomal enzyme system.

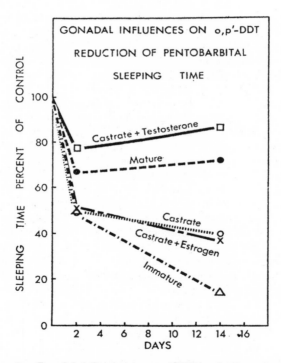

Figure 2. Gonadal influences on o,p'-DDT reduction of pento-barbital sleeping time. Each point represents five male rats fed 100 ppm o,p'-DDT in the diet. At 0 time, the immature, castrate, and castrate + estrogen slept 98, 143, and 188 min, respectively, while the castrate + testosterone and mature male rats slept 51 and 52 min. Sleeping times are expressed as percentages of these control values

130

If on the other hand the castrates were treated with testosterone, they reverted to the mature male rat state, and o,p'-DDT had less effect in reducing sleeping time in these testosterone-treated rats than in the castrates. The presence of the male sex hormone induces an elevated level of liver microsomal enzyme activity. Consequently, pesticide feeding is unable to stimulate liver microsomal induction further.

When the accumulation of pesticides in the rats was examined, we found that p,p'-DDT increased in an almost linear fashion in both immature male and female rats (Figure 3A) and in mature cycling or ovariectomized rats (Figure 3B). After 2 weeks feeding, pesticide concentrations were in the

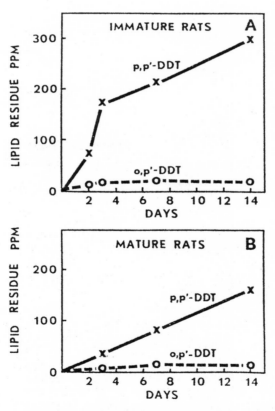

Figure 3. Accumulation of pesticides in rat lipid. A. Immature male and female rats. Each point represents ten rats. B. Mature ovariectomized and cycling female rats. Each point represents ten rats

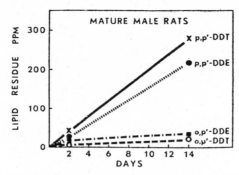

Figure 4. Accumulation of pesticides in rat lipid. Each point represents ten mature male rats fed 100 ppm of the pesticide

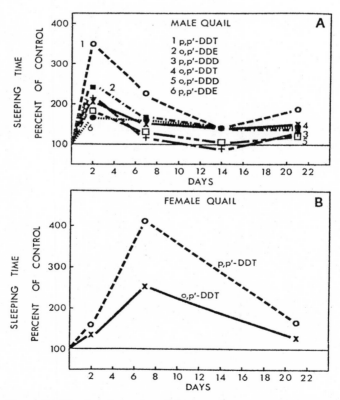

Figure 5. Sleeping times of quail fed pesticides. A. Male quail. Each point represents six quail. B. Female quail. Each point represents seven quail

range of 200–300 ppm in the adipose tissue of immature rats, and were about 150 ppm in the mature rats, showing an almost exact relationship between total intake of the pesticide and body weight of the rats. The o,p'-DDT residues in either immature or mature rats were only about $1/10$ of the

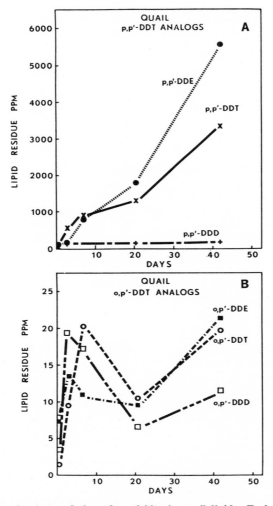

Figure 6. Accumulation of pesticides in quail lipid. Each point represents four quail

p,p'-DDT levels, being in the range of 15–20 μg/g in the lipid.

o,p'- and p,p'-DDE were also extremely effective in reducing sleeping time (Table I), and their accumulation in rat lipid is compared to that of the o,p'- and p,p'-isomers of DDT in Figure 4. A very similar pattern was seen, the p,p'-isomers accumulating and the o,p'-isomers being lost from the body.

In spite of this great disparity in the accumulation of the two isomers, indicating that the o,p'-compounds were almost completely metabolized and excreted from the body, the similar sleeping times observed with the two isomers indicated that the o,p'-analog was about as effective as p,p' in stimulating liver microsomal activity to degrade pentobarbital.

Table II. Toxicity to Pentobarbital: Percent Mortality

Days on Diet	Control	100 ppm o,p'-DDT	100 ppm p,p'-DDT
Male Quail			
2	9	64	82
6	17	0	33
14	0	17	50
Female Quail			
7	33	29	86

Average group size was seven quail.

These data suggest that toxicological effects in the case of this rapidly metabolized isomer are more closely related to intake than to body burden.

Figure 5 shows the effects of 3 weeks feeding of 100 ppm of o,p'- and p,p'-isomers of DDT, DDE, and DDD on pentobarbital sleeping time in quail. In complete contrast to the shortening of sleeping time observed in rats, the DDT isomers and metabolites prolonged sleeping time in quail. Two days after being placed on the feed, the male quail slept 1.5 to 3.5 times longer than untreated control quail in response to a standard dose of pentobarbital (Figure 5A). After 7 days the increase in sleeping time was not as great, and at 2 weeks the quail fed o,p'- and p,p'-DDD were not different from con-

134

trols in the sleeping time test. The quail fed DDT and DDE still showed an increased sleeping time about 150% of control, but the increases were considerably below the increases measured at 2 days.

The female quail (Figure 5B) also showed an increase in pentobarbital sleeping time, but the time course was somewhat different. Two days after being placed on DDT containing feed, sleeping times were only about 50% greater than controls. At 7 days, however, liver microsomal activity was inhibited to the extent that pentobarbital metabolism was 2.5 to 4 times less than that of controls, as judged by increased pentobarbital hypnosis. Again, this increase in sleeping time was not maintained, and after 3 weeks feeding, the sleeping time was about 50% higher than that of untreated control birds. The decline in sleeping time after an early increase, in both male and female quail, suggested that an adaptation was taking place in response to the chlorinated hydrocarbon stimulus. In order to determine whether the inhibition of liver metabolism of the barbiturate observed at 21 days would be maintained at longer times on a pesticide diet, sleeping times were determined on quail fed pesticide-containing diets for 6 weeks. At 42 days, quail fed 100 ppm of o,p'-DDT, p,p'-DDT, o,p'-DDE, or p,p'-DDE still showed an increased mean sleeping time of 153% of control.

An adaptation of this magnitude should markedly affect the acute toxicity of an administered drug. In male quail if the increase of sleeping time at 2 days reflects a lowered metabolism of pentobarbital, the acute toxicity to this drug should be different, and a dose which the control can tolerate might be reflected in a greater mortality in the pesticide treated group. Then, at later times when the adaptation takes place, this same lethal dosage should be tolerated by the body and lowered mortality should be observed.

Pentobarbital at a dose rate of 105 mg per kg of body weight was injected intramuscularly (Table II). This dose killed only 1 out of 11 control birds but was lethal to 64% and 82% of o,p'- and p,p'-DDT treated male quail. At 6 and 14 days the male quail metabolized this dose to a greater extent and showed a lower mortality than at 2 days. This same dosage was more lethal in female quail and killed about $1/3$ of the control females. Female quail fed p,p'-DDT for 7 days could not effectively metabolize the pentobarbital and 86% died in response to this dose. These mortality experiments thus also demonstrated the adaptation in liver metabolism induced by the pesticides.

In order to determine whether this sleeping-time adaptation pattern might be a reflection of differing accumulation levels of the pesticides, samples of lipid from quail were analyzed

135

for chlorinated hydrocarbon residues. p,p'-DDE and p,p'-DDT increased almost linearly with increasing time on the diet (Figure 6A). After 3 weeks on the diet, pesticide lipid concentrations were in the range of 1000–2000 ppm and after 42 days were in the 3000–5000 ppm range. p,p'-DDD rapidly increased to a level of about 100 ppm of lipid and remained in the 100–200 ppm range during the entire 42 day period. In marked contrast, the o,p'-isomers of DDT and the environmentally important analogs rapidly increased to concentrations of only 10–15 ppm in lipid and essentially remained at these levels during the entire feeding period (Figure 6B). Again, as observed previously in the rats, these extremely low accumulation levels of o,p'- analogs indicated that these compounds were almost completely metabolized and excreted from the body. In spite of this, however, they were about as effective as p,p'-compounds in influencing liver microsomal enzyme metabolism of pentobarbital. This again suggested that toxicological effects were more closely related to intake than to body burden.

The concentrations of the pesticides shown in Figures 2, 3, and 6 refer only to parent compounds that were fed to the individual groups. Each group had minor amounts of some of the other analogs present in the lipid residues. A summary of lipid accumulations for immature rats fed o,p'- or p,p'-DDT for 14 days and for quail fed for 42 days is presented in Table III. Both the rats and quail fed p,p'-DDT had the expected accumulation of p,p'-DDE and p,p'-DDD. In the case of o,p'-DDT feeding, both rats and quail showed minor concentrations of p,p'-DDE and significant concentrations of p,p'-DDT. These findings are in agreement with the report of Klein *et al.* (1964, 1965).

Klein *et al.* (1964, 1965) have interpreted their data to indicate that o,p'-DDT is converted to p,p'-DDT in the rat. However, in view of the large differences in the relative retentions of the two compounds in the animal body, we were concerned about the possibility of a minor contamination of the o,p'-DDT with p,p'-DDT. If the fractional retention of p,p'-DDT at low intakes is the same as at high intakes, the p,p'-DDT observed in rat lipid could be accounted for by a 0.48% contamination of the o,p'-DDT administered. The same value for quail would be 0.41%.

A recorder tracing of the o,p'-DDT at 1 μg/ml is shown in Figure 7A. In agreement with the findings of Klein *et al.* (1964), there was no evidence of p,p'-DDT. However, when a 100 μg/ml solution was used, several minor peaks were observed, including one with the same retention time as p,p'-DDT (Figure 7B). The identity was confirmed by tlc separation and subsequent recovery of the peak at the R_f corresponding to p,p'-DDT (Figure 7C). The tlc separation

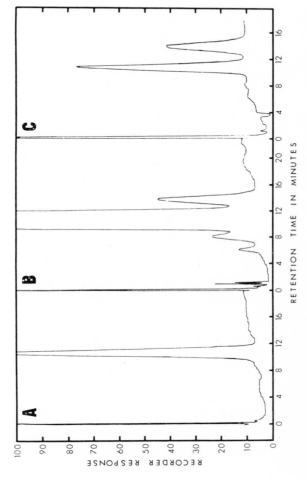

Figure 7. A. Glc recorder tracing of 3 μl of 1 μg/ml *o,p'*-DDT. B. Glc recorder tracing of 3 μl of 100 μg/ml of *o,p'*-DDT. The three minor peaks have retention times corresponding to *o,p'*-DDE, *p,p'*-DDE, *p,p'*-DDT, and *p,p'*-DDT, respectively. C. Glc recorder tracing of 3 μl of the material eluted from tlc at the R_f corresponding to *p,p'*-DDT. Volume of the eluate was equivalent to that required for a concentration of 100 μg/ml of the starting *o,p'*-DDT

137

established the presence of p,p'-DDT (Figure 7C), although the spot was not completely free of o,p'-DDT due to tailing of the starting material which contained over 99.5% o,p'-DDT.

It is difficult to precisely quantify the amount of p,p'-DDT contaminant in the commercially pure (99+%) preparation of o,p'-DDT that was fed. The average of several replicate determinations was 0.39 μg/ml in a 100 μg/ml solution of

Table III. Concentration of Analogs in the Lipid of Rats and Quail Fed o,p'-DDT or p,p'-DDT

Fed	Found[a]			
	p,p'-DDE	p,p'-DDD	p,p'-DDT	o,p'-DDT
			. μg/g	
Immature Rats[b]				
o,p'-DDT	1.15	...	1.33	21.7
p,p'-DDT	18.8	2.84	279	...
Quail[c]				
o,p'-DDT	1.93	...	13.7	19.4
p,p'-DDT	680	50.4	3310	...

[a] All values are an average of four animals. [b] Fed 100 μg/g diet for 14 days. [c] Fed 100 μg/g diet for 42 days.

o,p'-DDT. Thus, the animals fed 100 μg of o,p'-DDT/g diet would also receive 0.39 μg of p,p'-DDT/g diet. This value is in the same range as the amount of p,p'-DDT that would be required to produce our observed p,p'-DDT accumulations. Thus our results would not provide any evidence that o,p'-DDT is converted to p,p'-DDT in either rats or quail.

We have found no previous literature references to pesticide-sleeping time interactions in birds. In mice, Rosenberg and Coon (1958) have reported an increase of hexobarbital sleeping time when fed certain organophosphorus insecticides. Azarnoff *et al.* (1966) have reported that p,p'-DDD shortens hexobarbital sleeping time in rats and dogs but prolongs pentobarbital sleep in dogs. Rosenberg and Coon (1958) have suggested that the mechanism of this action was a competition for the enzyme systems which are responsible for destruction of hexobarbital.

There have been several previous investigations on the effects of pesticides in Japanese quail. Shellenberger *et al.* (1965, 1966) have studied the effects of several organophosphate pesticides on Japanese quail. Recently, Gillett and Arscott (1969) reported that DDT, fed at 100 ppm for several months, causes a depression of hepatic microsomal epoxidase activity in mature quail when aldrin and heptachlor were the substrates. In other experiments in which dieldrin was fed to young Japanese quail, aldrin epoxidase activity of hepatic microsomes was increased while cytochrome P-450 or NADPH-neotetrazolium reductase levels were not affected. Walker *et al.* (1969) have also studied the toxicity of dieldrin to Japanese quail. In studies on the effects of DDT and dieldrin on liver microsomal activity of hens, Sell *et al.* (1971) also found that the liver microsomal enzymes exhibited a marked independence. *p,p'*-DDT reduced liver microsomal aniline hydroxylase activity but did not significantly affect *N*-demethylase activity or cytochrome P-450 concentrations.

The concentrations of *p,p'*-DDT in the body fat of quail was approximately ten times that of rats (Figures 3, 4, and 6). This finding appears to be in conflict with older evidence summarized by Müller (1959), who concluded that "all species store DDT in their body fat at rates of the same order of magnitude when exposed repeatedly at the same dose rate." The apparent discrepancy is probably a result of two factors. We have found that quail consume more feed per unit body weight and have a lower body fat concentration than rats. As a result, the fraction of the consumed DDT retained does not differ greatly for the two species.

An interesting finding was the low amount of *o,p'*-isomers relative to *p,p'*-isomers in birds. In the rats, *o,p'*-isomers accumulated to about 15 ppm and *p,p'*-isomers were present at about 150 ppm, a ratio of about 1 to 10. In the quail, *o,p'*-isomers were also present at about 15 ppm in the lipid, but *p,p'*-concentrations were about 1500 ppm, a ratio of about 1 to 100. The fact that a similar proportion of the ingested *p,p'*-isomers is retained in both species suggests that the *o,p'*-isomers are more readily metabolized by the quail. These differences in the accumulation and metabolism of the persistent organochlorine pesticides in birds suggest a species difference and might fit well with the suggested greater adverse biological effects in avian as compared to mammalian species (Mrak, 1969).

LITERATURE CITED

Association of Official Agricultural Chemists, "Changes in Official Methods of Analysis," *J. Ass. Offic. Agr. Chem.* **49,** 227 (1966).
Azarnoff, D. L., Grady, H. J., Svoboda, D. J., *Biochem. Pharmacol.* **15,** 1985 (1966).
Bitman, J., Cecil, H. C., Harris, S. J., Fries, G. F., *Science* **162,** 371 (1968).
Bush, M. T., "Sedatives and Hypnotics. I. Absorption, Fate and Excretion," p. 206, *Physiol. Pharmacol.* Vol. I, W. S. Root and

F. G. Hofmann, Eds., Academic Press, New York, 1963.

Conney, A. H., Burns, J. J., *Advan. Pharmacol.* **1,** 31 (1962).

Conney, A. H., Welch, R. M., Kuntzman, R., Burns, J. J., *Clin. Pharmacol. Therap.* **8,** 2 (1967).

Gillett, J. W., Arscott, G. H., *Comp. Biochem. Physiol.* **30,** 589 (1969).

Hart, L. G., Shultice, R. W., Fouts, J. R., *Toxicol. Appl. Pharmacol.* **5,** 371 (1963).

Hart, L. G., Fouts, J. R., *Arch. Exp. Pathol. Pharmakol.* **249,** 486 (1965).

Klein, A. K., Laug, E. P., Datta, P. R., Watts, J. O., Chen, J. T., *J. Ass. Offic. Agr. Chem.* **47,** 1129 (1964).

Klein, A. K., Laug, E. P., Datta, P. R., Mendel, J. L., *J. Amer. Chem. Soc.* **87,** 2520 (1965).

Levin, W., Welch, R. M., Conney, A. H., *Fed. Proc.* **27,** 649 (1968).

Mrak, E. M., Report of the Secretary's Commission on Pesticides and Their Relationship to Environmental Health, U.S. Dept. HEW (1969).

Müller, P., "DDT, the Insecticide Dichlorodiphenyl Trichloroethane and its Significance," p. 68, Birkhauser, Basel, 1959.

Rosenberg, P., Coon, J. M., *Proc. Soc. Exp. Biol. Med.* **98,** 650 (1958).

Sell, J. L., Davison, K. L., Puyear, R. L., J. AGR. FOOD CHEM. **19,** (1971).

Shellenberger, T. E., Newell, G. W., Bridgman, R. M., Barbaccia, J., *Toxicol. Appl. Pharmacol.* **7,** 550 (1965).

Shellenberger, T. E., Newell, G. W., Adams, R. F., Barbaccia, J., *Toxicol. Appl. Pharmacol.* **8,** 22 (1966).

Walker, A. I. T., Neill, C. H., Stevenson, D. E., Robinson, J., *Toxicol. Appl. Pharmacol.* **15,** 69 (1969).

Welch, R. M., Levin, W., Conney, A. H., *Toxicol. Appl. Pharmacol.* **14,** 358 (1969).

DDT-Induced Inhibition of Avian Shell Gland
Carbonic Anhydrase: A Mechanism for Thin Eggshells

Joel Bitman
Helene C. Cecil
George F. Fries

The pesticide DDT (*1*) produces a decrease in eggshell thickness in Japanese quail (*2*), sparrow hawks (*3*), and mallards (*4*). The content of calcium in the eggshell declined (*2*) and reproduction was impaired (*3, 4*) by the direct addition of DDT or DDE (*1*) to the diet, thus confirming correlative evidence (*5, 6*) that DDT and related organochlorine compounds decrease eggshell thickness. We investigated carbonic anhydrase (CA) (E.C.-4.2.1.1) in the shell-forming gland of Japanese quail fed DDT or DDE to determine whether decreased activity could account for the defect in eggshell formation.

The Japanese quail were housed in individual cages on a schedule of 14 hours of light and 10 hours of dark. They were fed diets containing 100 ppm of *p,p'*-DDT or 100 ppm of *p,p'*-DDE for 3 months. Diets of both adequate (2.5 percent) and low (0.6 percent) calcium content were used. Activity of CA was assayed electrometrically by the Wilbur and Anderson procedure (*7*) as modified by Woodford *et al.* (*8*), in which substrate is supplied in the form of gaseous carbon dioxide. The saturated KCl solution of the combination electrode of the Corning model 12 *p*H meter was replaced with $4M$ KCl to prevent freezing out

141

Table 1. Carbonic anhydrase activity in the shell gland and blood of control quail and quail treated with DDT or DDE. Results are expressed as the number ± standard error.

Group	N	Shell gland weight (g)	CA activity		Blood (unit/ml)
			Shell gland		
			Unit/g	Total units	
Control	20	1.59 ± .09	186 ± 7	298 ± 22	1184 ± 96
p,p'-DDT	18	1.54 ± .08	156 ± 12*	242 ± 18†	924 ± 82†
p,p'-DDE	11	1.57 ± .07	150 ± 10*	235 ± 18†	663 ± 56‡

* $P < .01$.   † $P < .05$.   ‡ $P < .001$.

Table 2. Eggshell calcium and pesticide concentration in lipid and eggs of quail treated with DDE or DDT. Eggshell calcium is expressed as a percentage of egg weight.

Group	Eggshell calcium (%)	Eggs		Lipid	
		DDE (μg/g)	DDT (μg/g)	DDE (μg/g)	DDT (μg/g)
Control	2.58 ± .06	0.20	0.40	1.47	3.70
p,p'-DDT	2.37 ± .03*	48	196	483	1373
p,p'-DDE	2.38 ± .07†	196		1610	

* $P < .005$.   † $P < .05$.

142

of KCl in the asbestos fiber salt bridge and in the body of the electrode. All pH standardizations and reactions were conducted at 0°C. The CA standard was a purified preparation from beef blood (Worthington Biochemical Corp.).

Birds were killed by decapitation 6 to 8 hours before estimated oviposition, at which time a calcifying egg was present in the shell gland. Whole blood was collected in oxalated tubes. The samples of blood (0.5 to 1.0 ml) and weighed samples of whole shell gland (300 to 400 mg) were homogenized in ice-cold water for 3 minutes. The homogenates were centrifuged at $2500g$ for 15 minutes at 0°C. The supernatant solutions were then centrifuged again at $9500g$ for 20 minutes at 0°C. The opalescent supernatants (or a dilution) were then assayed immediately for CA activity. Calcium was determined by atomic absorption spectrophotometric analysis of solutions obtained by wet-ashing eggshell in concentrated HCl. Pesticide residues were determined in body fat and eggs by gas-liquid chromatography with an electron capture detector (9).

Carbonic anhydrase activity was lower in both the shell gland and blood of the treated Japanese quail (Table 1). Decreases of 16 to 19 percent occurred in CA from the shell gland of Japanese quail fed p,p'-DDT or p,p'-DDE. There were no differences in the weights of the shell glands among the groups. It was not possible to determine whether DDT and DDE caused a decrease in the total amount of enzyme or whether they partially inhibited the enzymatic activity in these extracts. The activity of CA in the blood of the quail treated with DDT or DDE exhibited larger declines—22 and 44 percent, respectively (Table 1).

The concentrations of pesticides in the lipid and eggs and the percent of calcium in the eggshells were determined (Table 2). Pesticide concentration in the eggs was approximately one-eighth of the body lipid concentration. Eggshell calcium was significantly lower in eggs from the quail treated with DDT or DDE.

Carbonic anhydrase is inhibited by DDT in human blood (10) and a sensitive method for DDT detection has been based on inhibition of bovine erythrocyte CA by DDT (11). In contrast, Anderson and March (12) were unable to demonstrate an effect of DDT on insect CA either in vivo or in vitro.

In the formation of the avian eggshell, CA is believed to be necessary to supply the carbonate ions required for calcium carbonate deposition. Several investigations have supported an active role for CA in eggshell formation (13), showing that CA was lower in shell glands producing soft-shelled eggs or no eggs than in glands producing normal eggs. Bernstein et al. (14) have provided additional evidence for an obligatory role for CA in eggshell formation.

Mueller (15) has questioned this role for CA because he did not find significant differences in CA activity in the shell gland at different stages of egg formation, suggesting that active shell formation was not accompanied by increased CA activity. Heald et al. (16) also did not find a significant correlation between CA activity and shell strength.

Treatment with DDT results in decreased CA activity in the avian shell gland. This demonstration in vitro does not preclude normal functioning of the CA enzymatic machinery in the intact tissue in vivo. Under the conditions of our experiments, however, the percentage declines in shell gland CA activity were 16 to 19 percent, amounts which could account for observed decreases in eggshell thickness of 10 to 15 percent in birds treated with DDT or DDE (2–4). The limitation by carbonic anhydrase of carbonate ions needed for the deposition of the calcium carbonate of the shell could provide the mechanism by which chlorinated hydrocarbons affect eggshell thickness.

143

References and Notes

1. Abbreviations: *p,p'*-DDT, 1,1.1-trichloro-2,2-bis(*p*-chlorophenyl)ethane; *p,p'*-DDE, 1,1-dichloro-2,2-bis(*p*-chlorophenyl)ethylene.
2. J. Bitman, H. C. Cecil, S. J. Harris, G. F. Fries, *Nature* **224**, 44 (1969).
3. R. D. Porter and S. N. Wiemeyer, *Science* **165**, 199 (1969).
4. R. G. Heath, J. W. Spann, J. F. Kreitzer, *Nature* **224**, 47 (1969).
5. D. A. Ratcliffe, *ibid.* **215**, 208 (1967).
6. J. J. Hickey and D. W. Anderson, *Science* **162**, 271 (1968).
7. K. M. Wilbur and N. G. Anderson, *J. Biol. Chem.* **176**, 147 (1948).
8. V. R. Woodford, N. Leegwater, S. M. Drance, *Can. J. Biochem. Physiol.* **39**, 287 (1961).
9. H. C. Barry, J. G. Hundley, L. Y. Johnson, Eds., *Pesticide Analytical Manual* (U.S. Dept. of Health. Education and Welfare, Food and Drug Administration, 1963; revised 1964 and 1965), vol. 1.
10. C. Torda and H. Wolff, *J. Pharmacol. Exp. Ther.* **95**, 444 (1949).
11. H. Keller, *Naturwissenschaften* **39**, 109 (1952).
12. A. D. Anderson and R. March, *Can. J. Zool.* **34**, 68 (1956).
13. R. H. Common, *J. Agr. Sci.* **31**, 412 (1941); M. S. Gutowska and C. A. Mitchell, *Poultry Sci.* **6**, 196 (1945).
14. R. S. Bernstein, T. Nevalainen, R. Schraer, H. Schraer, *Biochim. Biophys. Acta* **159**, 367 (1968).
15. W. J. Mueller, *Poultry Sci.* **41**, 1792 (1962).
16. P. J. Heald, D. Pohlman, E. G. Martin, *ibid.* **47**, 858 (1968).

144

DDT Intoxication in Birds: Subchronic Effects and Brain Residues

Elwood F. Hill, William E. Dale, and James W. Miles

Technical Development Laboratories, Laboratory Division, Center for Disease Control,
Health Services and Mental Health Administration, Public Health Service,
U.S. Department of Health, Education, and Welfare, Box 2167,
Savannah, Georgia 31402

DDT [1,1,1-trichloro-2,2-bis(*p*-chlorophenyl)ethane] is so widely distributed in nature that DDT and its principal metabolites DDD [1,1-dichloro-2,2-bis(*p*-chlorophenyl)-ethane], and DDE [1,1-dichloro-2,2-bis(*p*-chlorophenyl)ethylene] are commonly found in the tissues of almost all animals. The residue levels in brain tissue have been shown to be the best criteria for determining DDT poisoning in rats (*Rattus norvegicus*) (Dale *et al.*, 1963), house sparrows (*Passer domesticus*) (Bernard, 1963), and cowbirds (*Molothrus ater*) (Stickel *et al.*, 1966). Dale *et al.* (1962) demonstrated that when rats fed sublethal levels of DDT are subjected to partial starvation, DDT is mobilized from the fat depots into the blood stream and into the brain, with toxic signs resulting. This closely simulates the condition of stress in migrating birds. A similar experiment has been carried out in cockerels (*Gallus gallus*). It showed that food deprivation caused the mobilization of DDT from the fat depots. This condition (or mobilization) led to toxic signs and death (Ecobichon and Saschenbrecker, 1969).

145

To better establish the role of DDT as the cause of illness and death in wild birds in the field, the effects of various dietary levels of DDT on several avian species have been studied. In this study brain residues, weight changes, and other toxic effects are correlated to experimentally administered sublethal and lethal dietary levels of technical grade DDT in farm-reared bobwhite (*Colinus virginianus*). The results obtained on farm-reared bobwhite have been compared to those obtained on captive wild bobwhite, blue jays (*Cyanocitta cristata*), house sparrows, and cardinals (*Richmondena cardinalis*).

METHODS

Subadult farm-reared male bobwhite were randomized into study groups and acclimatized for 2 mo to the wire-mesh pens ($4 \times 6 \times 1.5$ ft high) and a diet of Pigeon Chow Checkers[2] (Ralston Purina Co., St. Louis, Missouri) before the treated feed was presented.

After acclimation dietary concentrations of 25, 50, 100, 200, 400, and 800 ppm technical grade DDT were presented to 10-bird groups for 5 days. A group of 12 birds received 1600 ppm and a group of 10 birds served as the control. The control birds were fed untreated feed. The rate at which DDT residue accumulated in the brain and the influence of the toxicant on body weight were compared among the groups. Observations for signs of intoxication were made.

The toxic concentrations were designed to kill about 50% of the birds on 1600 ppm and none at lower levels. It was intended that all levels would produce measurable brain residues of DDT and its metabolites, and that several would cause observable signs of intoxication. Since DDT is reported to degrade to DDD post-mortem in mouse tissue (Barker and Morrison 1964), the birds were processed immediately after death. When a bird died its weight was recorded. The brain was immediately removed, weighed (wet) and individually stored in 10% formalin until analyzed. After 5 days on a toxic diet the survivors were sacrificed with chloroform and processed as described above.

Dose-effect relationships for residue accumulation and weight change were analysed by analysis of variance (Snedecor and Cochran, 1967). The 5% level of significance was accepted. Brain residue accumulation curves were fitted by inspection.

Lethal Studies

The relative sensitivity of farm-reared and wild bobwhite, blue jays, house sparrows and cardinals to DDT, based on lethal effects, was determined by comparing the estimated 5-day median lethal concentrations (LC50), residue levels in brain tissue, and weight changes among the species.

The test protocol was similar to that of Heath and Stickel (1965) except that we used subadult birds and 5 geometrically spaced doses. A randomized group of each species was exposed to a randomly assigned concentration of technical DDT for 5 days followed by 3 days on nontoxic feed. Study group sizes were: cardinals, 3–5; blue jays, 3–5; wild bobwhite, 6; farm-reared bobwhite, 10; and house sparrows, 20. Normally the sex composition within each group was equal. Pre- and posttest weights and time to death were recorded. Brains from randomly selected birds dying during treatment were frozen

[2] The use of trade names is for identification purposes only and does not constitute endorsement by the Public Health Service or the U.S. Department of Health, Education, and Welfare.

(individually) until they were analyzed for residues. Body weights of the survivors were not compared because they were not weighed until after the study which included 3 days on untreated feed, and weight gains were likely.

Specimens were acclimated to captivity for 1–4 mo in facilities similar to those described for the sublethal bobwhite test. Birds failing to regain their capture weight were not tested. All species were fed Pigeon Chow Checkers. The pens for blue jays, house sparrows, and cardinals were $2 \times 2 \times 1.5$ ft high and were equipped with perches.

The 5-day LC50 values and the 95 % confidence limits were determined by the method of Litchfield and Wilcoxon (1949). Insecticide residues in the brain, and also body weight changes, were compared by analysis of variance, and the means were separated statistically to determine species differences. Logarithmic transformation of residue data was used to normalize the distributions and equalize the variances. Dose levels and sexes were combined for each species because house sparrow data showed no statistical difference in residue accumulation between the sexes or doses; the combined data for each species met Bartlett's test for homogeneity of variance (Snedecor and Cochran, 1967). The 5 % level of significance was accepted.

Feed Preparation

Seven and one-half pounds of the feed (screen to remove fine particles) was put in an 8 qt twin-shell, stainless steel blender (liquid–solids type). Technical grade DDT required to treat the feed at a given concentration was dissolved in 50 ml of acetone and put into a 1-liter oxygen breathing tank. A needle valve assembly, fitted with a stainless steel delivery tube ($14 \times 3/16$ in.) with a brass tip (0.01 in. orifice), sealed the tank. Freon-12, 500 ml, was forced into the oxygen tank from a pressure burette. The delivery tube was inserted into the trunnion of the blender. The blender was started, then the DDT was sprayed onto the feed as a mist by opening the needle valve. About 5 min was required to discharge the contents.

Samples of the spiked feed were extracted and analyzed by gas chromatography. Recoveries of 94–117 % ($\bar{X} = 102 \%$) of the desired concentrations were observed.

Brain Residue Analysis

Reagents. Except o,p'-DDE, analytical standards were obtained from the Pesticide Repository, Pesticides Research Laboratory, Perrine, Florida and were 99–100 % pure. The o,p'-DDE was prepared by the dehydrohalogenation of o,p'-DDT according to the method of Haller *et al.* (1945). After recrystallization from alcohol, its melting point was 78–78.5°C.

The *n*-hexane was redistilled over 200 ml of methyl alcohol plus 2 g of metallic sodium per 4000 ml of *n*-hexane. The first 1000 ml were discarded.

Apparatus. A MicroTek MT-220 gas chromatograph equipped with an electron capture detector, 130 mCi tritium source, was used for analysis. Operating conditions were: column temperature, 210°C; inlet and outlet blocks, 250°C; detector, 200°C; carrier gas, nitrogen, 60 ml/min at 70 psi; background current, 5.12×10^{-9} amp; column I, aluminum tube $5' \times 1/4$ in., packed with 1.5 % OV-17 plus 2.5 % OV-210 on 100/120 mesh Chromosorb W, A.W., DMCS-treated; column II, aluminum tube $10' \times 3/16$ in., packed with 3 % OV-225 on 100/120 mesh Chromosorb W, A.W., DMCS-treated.

Procedure. The brains were blotted dry with white absorbent paper, weighed, and then ground in a mortar with the aid of 2–3 g of sea sand. The ground samples were put in Teflon-lined, screw-capped bottles and extracted with *n*-hexane for 1 hr on a wrist-action shaking machine.

Brain extracts from birds on 800 ppm DDT or more were analyzed without clean-up by injecting appropriate aliquots directly into the gas chromatograph equipped with

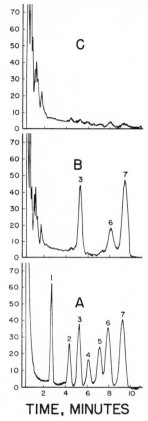

TIME, MINUTES

FIG. 1. Chromatograms (A) Standard pesticide mixture, peaks representing: *1*, Aldrin = 0.1 ng; *2, o,p'*-DDE = 0.08 ng; *3, p,p'*-DDE = 0.08 ng; *4, o,p'*-DDD = 0.1 ng; *5, o,p'*-DDT = 0.14 ng; *6, p,p'*-DDD = 0.17 ng; and *7, p,p'*-DDT = 0.32 ng; (B) Aliquot of extract equivalent to 0.2 mg of brain of bobwhite fed 400 ppm DDT in diet; (C) Aliquot of extract equivalent to 0.2 mg of brain of control bobwhite.

column I described above. Extracts from those dosed at lower levels had to be cleaned up. The following procedure was used: a 10-ml aliquot of the extract was pipetted into a 15-ml centrifuge tube and evaporated to near dryness in a 90°C water bath. Tube walls were then rinsed with 1 ml of *n*-hexane to concentrate the insecticide in the bottom of the tube. The solvent was then evaporated to about 0.5 ml. The samples were cleaned up as described by Enos *et al.* (1967). After clean-up the samples were analyzed for DDT and its metabolites by injecting aliquots into column I of the gas chromatograph operated as described. A typical chromatogram obtained with column I is presented in Fig. 1. Column II was used for qualitative confirmation.

To determine the efficiency of the method, *p,p'*-DDE, *o,p'*-DDT, *p,p'*-DDD, and *p,p'*-DDT were added to control brains at various levels and the extracts were carried through the described procedure. Excellent recoveries (96–105%) were obtained for each compound (Table 1).

TABLE 1

RECOVERY OF ADDED DDD, DDE, AND DDT FROM BOBWHITE BRAIN TISSUE BY GAS CHROMATOGRAPHY

Compound	Wet weight (ppm)		Percent recovered
	Added	Recovered	
p,p'-DDE	2.23	2.25	101
	4.46	4.41	99
	11.15	11.03	99
o,p'-DDT	4.04	4.24	105
	8.08	8.08	100
	20.20	19.80	98
p,p'-DDD	5.51	5.51	100
	11.02	10.79	98
	27.55	26.45	96
p,p'-DDT	5.42	5.69	105
	10.83	10.72	99
	27.08	25.99	96

RESULTS

Sublethal Studies

Data showing the accumulation rate of *p,p'*-DDE, *o,p'*-DDT, *p,p'*-DDD, and *p,p'*-DDT in bobwhite brains after 5 days of sublethal dietary exposure to DDT are summarized in Table 2. The toxic concentrations were spaced geometrically (2×), but the mean total brain residue accumulation differed between progressive levels as follows: 25–50 ppm, 1.4 X; 50–100 ppm, 1.8 X; 100–200 ppm, 1.9 X; 200–400 ppm, 4.5 X; and 400–800 ppm, 1.6 X. Individually all residues, accumulated similarly except *o,p'*-DDT. This may have been due to isomeric conversion of the *o,p'*-isomer, as French and Jefferies (1969) suggested, or to a faster excretion rate or metabolism to some undetected metabolite.

Figure 2 shows the brain residue accumulation curve for DDD + DDT based on 25–800 ppm DDT in the diet. DDD and DDT are combined because the combination is considered diagnostic of lethality (Stickel *et al.*, 1966). Seven birds out of 12 on 1600 ppm died, therefore residue values for this group were not plotted as part of the curve but were shown in comparison to the projection. Survivors of 1600 ppm had DDD + DDT residues that approximated the projected curve at 1600 ppm, whereas those dying were considerably above it (Fig. 2). Bobwhite dying on the 1600 ppm diet averaged 1.9 X more DDD + DDT than their surviving counterparts.

TABLE 2

BOBWHITE BRAIN RESIDUES (PPM, WET WEIGHT) AFTER 5 DAYS EXPOSURE TO SUBLETHAL DIETARY CONCENTRATIONS OF TECHNICAL DDT

Dietary concentration (ppm)	Number of samples	Evaluation	p,p'-DDE	p,p'-DDD	o,p'-DDT	p,p'-DDT	Total
0	10	Range	0.01–0.05	0.01–0.02	0.02–0.07	0.01–0.09	0.05–0.21
		Mean (SE)	0.02 (0.004)	0.01 (0.006)	0.04 (0.005)	0.06 (0.007)	0.13 (0.025)
25	10	Range	0.08–0.13	0.02–0.04	0.01–0.07	0.18–0.28	0.32–0.50
		Mean (SE)	0.10 (0.005)	0.03 (0.003)	0.05 (0.006)	0.22 (0.012)	0.40 (0.022)
50	9	Range	0.09–0.19	0.03–0.05	0.04–0.07	0.28–0.45	0.46–0.74
		Mean (SE)	0.13 (0.011)	0.03 (0.002)	0.06 (0.003)	0.34 (0.023)	0.57 (0.033)
100	10	Range	0.10–0.38	0.03–0.12	0.03–0.08	0.40–1.21	0.65–1.79
		Mean (SE)	0.21 (0.025)	0.07 (0.008)	0.06 (0.004)	0.69 (0.076)	1.02 (0.108)
200	10	Range	0.14–0.66	0.06–0.19	0.08–0.38	0.88–2.15	1.18–3.09
		Mean (SE)	0.34 (0.059)	0.12 (0.015)	0.13 (0.028)	1.35 (0.158)	1.94 (0.221)
400	10	Range	0.43–3.03	0.08–1.36	0.08–0.25	0.69–12.59	1.10–16.73
		Mean (SE)	1.36 (0.301)	0.55 (0.128)	0.13 (0.017)	6.71 (1.111)	8.75 (1.512)
800	10	Range	1.16–5.24	0.74–2.36	—	5.58–22.26	7.54–29.59
		Mean (SE)	2.77 (0.465)	1.35 (0.172)	—	10.08 (1.534)	14.20 (2.083)

150

Bobwhite weight changes after 5 days on 25, 50, 100, and 200 ppm could not be statistically separated from the controls. Groups given 400 ($n = 10$, $\bar{x}$ wt loss = 3%) and 800 ppm ($n = 10$, $\bar{x}$ wt loss = 2%) lost more weight than the controls and those on 25–200 ppm ($P < 0.01$), but they lost less than the survivors of the 1600 ppm group ($P < 0.01$). Birds dying on the 1600 ppm diet ($n = 7$, $\bar{x}$ wt loss = 12%) lost more weight than the survivors ($P < 0.05$) maintained on the same diet.

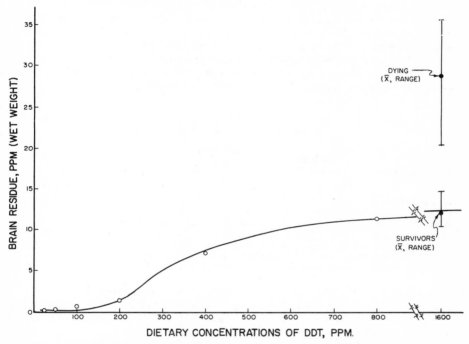

Fig. 2. DDD + DDT in the brain of bobwhite exposed to various sublethal dietary concentrations of DDT (25–800 ppm) compared to residues causing death and to survivors of 1600 ppm DDT.

No signs of intoxication were observed in the 25–200 ppm groups. The maximum brain residues of DDT plus metabolites associated with these groups was 3.1 ppm. At 400 ppm 9 of 10 specimens lost weight, but there were no observable toxic signs.

All birds dosed at 800 ppm displayed mild toxic signs after about 80 hr as manifested by slight tail tremors and occasional irregularities in head carriage. The severity of the signs intensified by the end of the 5 days. The minimum total residue associated with observable signs of intoxication was 7.5 ppm (DDD + DDT, 6.4 ppm).

Signs of poisoning were distinct in most birds on 1600 ppm after 30 hr and rapidly accentuated until they could be classed as "moderate" (conspicuous tail tremors, stumbling gait, occasional efforts to balance with aid of the wings, and head bobbing). The severity of the tremors intensified in 7 of the 12 specimens, leading to convulsions and death. The first and last birds died after 70 and 100 hr, respectively. The average time to death was 88.9 (SD = 11.6) hr. The severity of the signs in the 5 survivors appeared to have regressed to a point between "mild" and "moderate" by the time the studies were concluded.

Lethal Studies

The toxicity of technical DDT, as determined by the estimated median lethal concentration (LC50) to various avian species is presented in Table 3. Except for blue jays, the sensitivity was inversely related to body weight. Blue jays and house sparrows, comparable in sensitivity, were about 1.3× as sensitive as cardinals, about 2.2× as sensitive as wild bobwhite and 3× as sensitive as farm-reared bobwhite. Wild bobwhite were about 1.4× more sensitive than the farm variety, but their 95% confidence limits overlap. Confidence limits were not determined for blue jays and cardinals because of the small numbers in the study groups. No control birds died during the studies.

TABLE 3

Toxicity of DDT to Four Avian Species

Species	Birds per concentration	Pretest weight (g)		LC50[a] (ppm)	95% Conf. limits
		Range	Mean		
Blue jay	3–5	56–85	72.8	415	320–540[b]
House sparrow	20	23–31	26.7	415	370–465
Cardinal	3–5	30–47	37.9	535	420–700[b]
Bobwhite (wild)	6	126–158	143.8	1170	830–1650
Bobwhite (farm)	10	181–228	202.4	1610	1331–1948

[a] LC50's are based on 5 days of ad libitum toxic diet followed by 3 days of nontoxic feed.
[b] Highest concentration failing to kill 50% and lowest killing more than 50%.

Residues of DDD, DDE, and DDT in avian brain tissue after death from DDT poisoning are shown in Table 4. House sparrows generally had the highest residue and blue jays the lowest. DDT accumulated more uniformly among the species than did its metabolites. Even though some means were statistically separable, all ranges overlapped. DDE levels were significantly greater in house sparrows and bobwhite than in cardinals and blue jays. Although DDE tends to increase with the length of time of exposure to DDT, no correlation between time and residues was demonstrated in this study, probably because of its short duration. Significantly more DDD was found in house sparrows than in any other species, and wild bobwhite had more than farm-reared bobwhite ($P < 0.01$). Combined DDD and DDT was higher in house sparrows than in the other species ($P < 0.01$). There was no significant difference in the levels of DDD + DDT found in the brains of bobwhite and cardinals. However, residues were significantly higher in wild bobwhite than in blue jays ($P < 0.01$).

In dying birds, the minimum DDD + DDT brain residue levels for each dietary concentration were almost always highest in house sparrows. Minimum DDE residues were highest in wild bobwhite. There was no apparent relationship between minimum residues and either time to death or exposure levels for any species. The minimum residue levels were relatively constant within each species and were independent of dietary concentration.

TABLE 4

RESIDUES (PPM WET WEIGHT) IN BRAINS OF BIRDS DYING ON AD LIBITUM DIETS CONTAINING TECHNICAL DDT

Dietary concentration (ppm)	Number of samples	p,p'-DDE		p,p'-DDD		p,p'-DDT		p,p'-DDD + p,p'-DDT	
		Range	Geom. $\bar{x}$	Range	Geom. $\bar{x}$	Range	Geom. $\bar{x}$	Range	Geom. $\bar{x}$
House Sparrow									
320, 420, 540, 700	19	4.5-18.3	9.3	8.0-28.9	15.6	18.0-37.9	27.6	32.6-62.6	43.2[a]
Bobwhite (wild)									
1540, 2000	8	8.8-13.3	10.5	5.6-13.8	8.3	16.5-28.6	22.9	24.7-42.4	31.3[b]
Bobwhite (farm)									
1600	7	7.9-11.0	9.2	1.7-3.6	2.8	18.7-32.4	25.4	20.4-35.8	28.4
Cardinal									
420, 540, 700, 910	4	2.3-3.0	2.6	5.5-9.8	7.8	16.6-24.0	18.8	22.7-33.8	26.7
Blue Jay									
320, 540, 910	5	2.0-4.4	3.0	5.9-8.6	7.0	12.0-19.6	16.0	17.9-28.2	23.0

[a] Significantly greater than for other species ($P < 0.01$).
[b] Significantly greater than for blue jays ($P < 0.01$).

TABLE 5

Bobwhite and House Sparrow Weight Changes after Death from Technical DDT[a]

Dietary concentration (ppm)	Number of samples	Body weight (g)				Percent weight change[b]	
		Pretreatment		Death			
		Mean (SE)	Range	Mean (SE)	Range	Range	Mean
House Sparrow[c]							
320	4	27.2 (0.36)	26–29	24.5 (0.47)	24–25	−14.0 to −7.7	−9.9
420	11	26.5 (0.64)	23–29	23.8 (0.55)	21–26	−13.8 to −7.1	−10.2
540	15	26.7 (0.51)	23–31	23.6 (0.39)	21–27	−18.5 to −4.0	−11.6
700	18	27.3 (0.25)	26–29	23.8 (0.20)	22–25	−17.9 to −3.8	−12.8
Bobwhite (wild)[c]							
1540	4	142.8 (1.90)	134–149	114.5 (5.78)	103–125	−24.3 to −16.1	−19.8
2000	4	142.5 (4.16)	131–154	113.0 (4.32)	106–124	−24.1 to −18.5	−20.7
Bobwhite (farm)[c]							
1540	3	193.3 (6.57)	184–206	168.3 (6.69)	164–181	−13.7 to −12.1	−13.0
2000	4	201.5 (9.00)	186–220	172.8 (6.85)	158–188	−15.1 to −13.2	−14.1

[a] Toxicant was provided for 5 days; time to death ranged from 2 to 6 days.
[b] Mean weight loss at death was significant for all groups ($P < 0.01$).
[c] Mean death weights on the various concentrations were not statistically different.

154

The magnitude of weight changes for bobwhite (farm-reared and wild) and house sparrows dying from DDT is demonstrated in Table 5. Weight loss at death was significant in all instances. Weight loss within a species was similar regardless of exposure level. Mean weights at death could not be statistically separated by toxic level for either species. No correlation was established between pretreatment weight and numbers dying or time to death. There was no difference between the sexes in relation to weight or mortality.

Signs of intoxication for all species were similar to those reported for farm-reared bobwhite on sublethal dosage except for the passerines which appeared to have more difficulty walking. In all cases, the onset of observable signs was marked by relatively inconspicuous tail tremors.

DISCUSSION

Dose-response curves for brain residues of DDD, DDE, and DDT in bobwhite exposed to sublethal concentrations of technical DDT were obtained. The residue accumulation increased progressively between toxic levels up to 400 ppm; the greatest increase occurred between 200 and 400 ppm. This disproportionate increase (2.4× greater than between 100 and 200 ppm) probably indicates the range within which dietary DDT concentrations began to overwhelm the metabolic and excretory capability of this species during the 5-day test. Other than significant weight losses, gross signs of intoxication were absent at the 400 ppm level.

On the 800 ppm diet, observable signs of intoxication, significant weight loss and relatively high brain residues occurred. As with the sub-400 ppm levels, the variation in brain residues was small (3.2×), indicating uniformity in response. Weight losses, high brain residues, and an acceleration in signs of intoxication as the study progressed suggest that prolonged exposure could not be tolerated. One bird, although surviving the 5 days, accumulated potentially lethal brain residues (as compared to residues that were lethal at 1600 ppm).

Residues in birds dying on the 1600 ppm diet were clearly greater than those in the survivors. The lower extreme in the dead was about 1.5× the upper extreme in the survivors. Residue accumulation within each of these groups was reasonably uniform; those dying had an upper extreme of only 1.7× the lower, and the survivors had a spread of 1.5×.

Signs of intoxication in birds that died on 1600 ppm progressed rapidly through varying degrees of tremors. The tremors were associated with increasing ataxia and often culminated in convulsions. In contrast, the severity of signs in survivors appeared to peak with moderate tremors after 90–100 hr then regress. Apparently at least temporary adjustment to this potentially lethal dietary concentration of DDT was accomplished by some individuals. Cranmer et al. (1970) reported residue reduction in squirrel monkeys (*Saimiri* sp.) during repeated dosing.

The minimum brain residue associated with signs of intoxication was 7.5 ppm (800 ppm group). Overlapping occurred between residues in birds that were exposed to 400 ppm that did not show signs of intoxication and those on 800 ppm that did show signs. These differences may be due to residues accumulating so rapidly at 800 ppm that the birds had less time to compensate physiologically. It is evident that observable nervous system involvement is not initiated at the same brain residue levels in all

155

individuals. For wild birds, this apparently mild neurological disorder associated with low brain residue levels could present a handicap to survival.

Regardless of the dietary concentration causing death, there was marked uniformity in DDD + DDT brain residues within each species tested. Comparison of residues associated with death in house sparrows at 4 dietary concentrations revealed no statistical differences. Minimum DDD + DDT residues from the different lethal dietary concentrations failed to relate the variation in residues to dietary concentration for any species.

Death from DDT poisoning has been correlated to brain residues, but there is no clear-cut line between nonlethal and lethal amounts. Using data from various bird studies Stickel *et al.* (1966) proposed 30 ppm of DDD + DDT in the brain as indicative of death or serious danger. Except for house sparrows, our results for all species were generally lower than 30 ppm and suggest that 20 ppm would be critical, especially since we observed signs of intoxication associated with less than 10 ppm.

The apparent hypersensitivity of blue jays was surprising since Bernard and Wallace (1967) considered them to be comparatively tolerant of DDT on the basis of field studies in Michigan. Initially the small study group sizes were suspect in the low LC50 estimate of 415 ppm, but the low brain residues associated with mortality strengthened the plausibility of the result.

The LC50 for house sparrows was the lowest (415 ppm) of the species tested, but the brain residues at death were the highest. Wild bobwhite, on the other hand, had much lower brain residues but withstood higher dietary concentrations (LC50, 1170 ppm). This difference in response may be due to the higher metabolic rate of the smaller house sparrow which would require proportionally greater food consumption, and cause ingestion of more toxicant per unit of body size. Further, animals of high basal metabolism have reduced fat stores and the DDT would more readily translocate through the blood to vital organs.

ACKNOWLEDGMENTS

C. Wayne Thaggard, of our laboratory, assisted in all aspects of the avian studies including specimen procurement, maintenance, and testing. Mrs. Sue M. Hill, Bowie, Maryland, assisted with dissection and material preservation. Robert G. Heath, Patuxent Wildlife Research Center, Laurel, Maryland, advised on the presentation of certain data. Lee M. Alderman, Technical Development Laboratories, Savannah, Georgia, assisted in analyzing the samples. William H. Stickel, Patuxent Wildlife Research Center, critically reviewed the manuscript.

REFERENCES

BARKER, P. S., and MORRISON, F. O. (1964). Breakdown of DDT to DDD in mouse tissue. *Can. J. Zool.* **42**, 324–325.

BERNARD, R. F. (1963). Studies on the effects of DDT on birds. *Mich. Univ., Publ. Mus. Biol. Ser.* **2**, 155–192.

BERNARD, R. F., and WALLACE, G. J. (1967). DDT in Michigan birds. *Jack-Pine Warbler* **45**, 11–17.

CRANMER, M., COPELAND, F., and CARROLL, J. (1970). Dynamics of DDT storage in the squirrel monkey. *Toxicol. Appl. Pharmacol.* **16**, 9 (Abstract).

DALE, W. E., GAINES, T. B., and HAYES, W. J., JR. (1962). Storage and excretion of DDT in starved rats. *Toxicol. Appl. Pharmacol.* **4**, 89–106.

DALE, W. E., GAINES, T. B., HAYES, W. J., JR., and PEARCE, G. W. (1963). Poisoning by DDT: Relation between clinical signs and concentration in rat brain. *Science* **142**, 1474–1476.

ECOBICHON, D. J., and SASCHENBRECKER, P. W. (1969). The redistribution of stored DDT in cockerels under the influence of food deprivation. *Toxicol. Appl. Pharmacol.* **15**, 420–432.

ENOS, H. F., BIROS, F. J., GARDNER, D. T., and WOOD, J. P. (1967). Micro modification of the Mills procedure for the determination of pesticides in human tissue. Chicago A.C.S. Meeting, Div. Agr. Food Chem., Sept. 1967.

FRENCH, M. C., and JEFFERIES, D. J. (1969). Degradation and disappearance of ortho, para isomer of technical DDT in living and dead avian tissues. *Science* **165**, 914–916.

HALLER, H. L., BARTLETT, P. D., DRAKE, N. L., NEWMAN, M. S., CRISTAL, S. J., EAKER, C. M., HAYES, R. A., KILMER, G. W., MAGERLEIN, B., MUELLER, G. P., SCHNEIDER, A., and WHEATLEY, W. (1945). The chemical composition of technical DDT. *J. Amer. Chem. Soc.* **67**, 1599.

HEATH, R. G., and STICKEL, L. F. (1965). Protocol for testing the acute and relative toxicity of pesticides to penned birds. *In* "The effects of pesticides on fish and wildlife." *U.S. Fish Wildl. Serv. Circ.* **226**, 18–21.

LITCHFIELD, J. T., and WILCOXON, F. (1949). A simplified method of evaluating dose-effect experiments. *J. Pharmacol. Exp. Ther.* **96**, 99–113.

SNEDECOR, G. W., and COCHRAN, W. G. (1967). *Statistical Methods*, 6th ed. Iowa State Univ. Press, Ames, Iowa.

STICKEL, L. F., STICKEL, W. H., and CHRISTENSEN, R. (1966). Residues of DDT in brains and bodies of birds that died on dosage and in survivors. *Science* **151**, 1549–1551.

157

DDT in Lower Vertebrates

Influence of DDT upon Pituitary Melanocyte-Stimulating Hormone (MSH) Activity in the Anuran Tadpole

Anuran tadpoles exposed to DDT at sublethal concentrations were found to have significantly increased pituitary melanocyte-stimulating hormone levels when compared with control animals. An effect of DDT upon the hypothalamus is postulated.

Attempts have been made to link melanocyte-stimulating hormone (MSH) with various aspects of mammalian physiology. Evidence produced thus far lends support to the hypothesis of a neurotropic activity. A review of the extrapigmentary effects of MSH, including its effect upon monosynaptic potentials and the induction of EEG changes of low wakefulness by injection of MSH, has been compiled by Kastin et al. (1968). Behavioral responses have been demonstrated by Ferrari (1963), De Wied and Bohus (1966) and Sandman et al. (1969). Recently the direct relationship between the tranquilizing drugs and anesthetics and the plasma and pituitary levels of MSH demonstrated by Kastin et al. (1969) would seem to further strengthen the theory.

In view of the apparent relationship between DDT (1,1,1-trichloro-2,2-di-p-chlorophenylethane) and the nervous system, specifically upon a Na^+, K^+, Mg^{2+}-adenosine triphosphatase (Matsumura and Patil, 1969) and the tendency of DDT to bind with synaptic junctions (Brunnert and Matsumura, 1969), it seemed reasonable to investigate the effect of DDT upon MSH levels.

Rana clamitans larvae of the forelimb bud stage were exposed to DDT by the addition of varying amounts of a 50% wettable powder of this compound directly to the aquaria. A ratio of one tadpole to 1 liter of water was maintained with the concentration of DDT ranging from 0.1–0.8 ppm. Tanks were kept on a black background and both test and control animals were exposed to normal daylight conditions. Tadpoles at DDT concentrations of 0.5–0.8 ppm were autopsied after 24 hr ex-

posure; those at concentrations of 0.1–0.5 ppm were autopsied after 6 days. Pituitaries were removed in midafternoon after chloroform anesthesia and MSH levels were determined using the *in vitro* frog skin assay of Lerner and Wright (1960) with the modifications introduced in our laboratory (Peaslee et al., 1969). Results are expressed in units of MSH activity/whole pituitary and are found in Table 1. Percent change of MSH activity in each case is based on the control animals of the same group. Animals exposed to the higher concentration were somewhat less developed and this may account for the lower control value for MSH. The pituitaries of animals exposed to DDT in each case showed a significantly increased level of MSH activity.

In view of the evidence supporting an inhibition of MSH release under hypothalamic control, MSH-release inhibiting

TABLE 1
LEVEL OF MSH ACTIVITY IN PITUITARIES OF TADPOLES EXPOSED TO DDT

Concentration of DDT (ppm)	No. of animals	MSH activity U/pituitary[a]	% Change
0.1–0.5			
Group 1			
Test	5	284 ± 41	+59
Control	3	179 ± 30	—
Group 2			
Test	5	62 ± 11	+88
Control	5	33 ± 8	—
0.5–0.8			
Test	6	183 ± 20	+129
Control	2	80 ± 32	—

[a] Results are expressed as mean ± SE.

factor (MIF) (Kastin *et al.*, 1969), it seems reasonable that DDT could exert influence over this neuroendocrine system. Control and test animals exhibited a similar appearance indicating that the amount of circulating MSH was approximately the same in each case. Further studies are necessary to determine whether this elevated pituitary MSH activity is due to increased release inhibition (elevated MIF) or an increased synthesis of the hormone itself. Since there is no assurance that the melanocyte-stimulating activity exhibited is due to MSH rather than ACTH, more detailed examinations of the influence of DDT upon the hypothalamic-hypophyseal complex are being carried out.

REFERENCES

Brunnert, H., and Matsumura, F. (1969). Binding of 1,1,1-Trichloro-2,2-di-*p*-chlorophenylethane (DDT) with subcellular fractions of rat brain. *Biochem. J.* 144, 135–139.

De Wied, D., and Bohus, B. (1966). Long term and short term effects on retention of a conditioned avoidance response in rats by treatment with long acting pitressin and alpha-MSH. *Nature* 212, 1484–1486.

Ferrari, W., Gessa, G. L., and Vargui, L. (1963). Behavioral effects induced by intracisternally injected ACTH and MSH. *Ann. N. Y. Acad. Sci.* 104, 330–345.

Kastin, A. J., Kullander, S., Borglin, N. E., Dahlberg, B., Dyster-Aas, K., Krakau, C. F., T., Ingvar, D. H., Miller, M. C., Bowers, C. Y., and Schally, A. V. (1968). Extrapigmentary effects of melanocyte-stimulating hormone in amenorrhoeic women. *Lancet* 1, 1007–1010.

Kastin, A. J., Schally, A. V., Viosca, S., and Miller, M. C. (1969). MSH activity in plasma and pituitaries of rats after various treatments. *Endocrinology* 84, 20–27.

Lerner, A. B., and Wright, M. R. (1960). *In vitro* frog skin assay for agents that darken and lighten melanocytes. *In* "Methods of biochemical analysis" (D. Glick, ed.), vol. 8, pp. 295–307. Interscience. New York.

Matsumura, F., and Patil, K. C. (1969). Adenosine triphosphatase sensitive to DDT in synapses of rat brain. *Science* 166, 121–122.

Peaslee, M. H., Goldman, M., and Naber, S. P. (1969). Influence of melatonin upon melanocyte-stimulating hormone (MSH) activity in weanling and young adult rats. *Proc. S. Dak. Acad. Sci.* 48, 39–43.

Sandman, C. A., Kastin, A. J., and Schally, A. V. (1969). Melanocyte-stimulating hormone and learned appetitive behavior. *Experientia* 25, 1001–1002.

Margaret H. Peaslee

JAMES L. COX

Accumulation of DDT Residues in *Triphoturus mexicanus* from the Gulf of California

CONTAMINATION of marine fish by chlorinated hydrocarbons, especially by DDT and its congeners*, could threaten their future or continued utility as a food source if residues accumulate to the point of incipient toxicity or detrimental sublethal effects[1]. Little is known about the distribution of DDT residues in marine fish beyond listed concentration values for certain species. Most investigations have dealt with concentrations of residues in tissues or large pooled, unsorted samples of commercially caught fish (refs. 3 and 4: the latter covers exclusively marine studies). From this limited information, we know that fish of a single species caught in adjacent areas have markedly different contents of residues, probably because of differences in the magnitude of local sources of estuarine or airborne pesticides[2-5]. This heterogeneity of exposure poses problems for the interpretation of residue data from fish caught in these areas. In an attempt to obtain size-class data about concentrations of residues, relatively free from the effects of pesticide "hot spots"[6], *Triphoturus mexicanus*, a midwater fish from an area relatively remote from areas of pesticide application, was chosen for analysis.

Samples of midwater fish were collected in a six foot Tucker trawl at several locations in the Gulf of California. Fish were immediately deep-frozen for later analysis. All fish were later thawed and weighed, then sorted and pooled in groups of narrow size range for processing. The samples were digested in a mixture of acetic and perchloric acids, and the lipid fraction containing the pesticide was extracted from the diluted digestion mixture with very pure n-hexane[7]. Lipids and interfering coextractives were removed by passage of the extract through an acid–'Celite' column[7]. No further cleaning was necessary. The extracts were concentrated and injected into a Beckman GC-4 gas chromatograph with two columns and two electron-capture detectors. Each sample was chromatographed on at least two columns; columns used were 5 per cent QF-1, 5 per cent DC-200, or a mixed bed of both column materials, all on DCMU 'Chromosorb W'. Operating parameters were those recommended by the US Food and Drug Administration[8]. Confirmatory

---

* Metabolites of p,p'-DDT, o,p'-DDT and other constituents of technical DDT; only p,p'-DDT, p,p'-DDD, and p,p'-DDE were detected in concentrations greater than trace values in the analyses.

Table 1. DDT RESIDUES IN *Triphoturus mexicanus* FROM THE GULF OF CALIFORNIA

Station	Latitude	Longitude	No. of samples	Total No. of individuals	Weight range (mg)	Parts/10^9 ($\pm$ s.e.) Total conc.	DDE	Percentage of DDT	DDD
55	24° 13′ N	109° 22′ W	1	1	996	14	36	36	28
56	24° 13′ N	109° 18.5′ W	1	1	2,483	26	27	62	11
84	27° 40′ N	111° 22.5′ W	4	21	29–711	31($\pm$10)	20	73	7
85	27° 33.5′ N	111° 27.5′ W	9	23	29–1,196	37($\pm$10)	35	52	13
87	28° 07′ N	112° 26′ W	1	5	378	46	15	76	9
118	28° 25′ N	112° 30′ W	1	2	635	58	10	85	5
*62	27° 14.5′ N	111° 58′ W	4	64	21–867	45($\pm$24)	23	64	13

* Sample from cruise 16 of Stanford oceanographic expeditions aboard the RV Te Vega during the summer of 1967. All other samples were collected during Stanford oceanographic expedition 21, in August 1969, from RV Proteus.

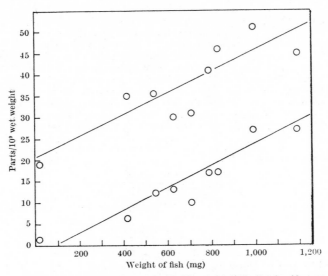

Fig. 1. Plot of sample weight against residues (parts/10^9). Total residues include all those DDT compounds which were detectable. Each point represents pooled samples of several fish. Weight ranges within the pooled samples was in most cases less than 10 per cent of the mean weight for the group. Top line, total of residues analysed (p,p'-DDT, DDE, DDD); bottom line, p,p'-DDE only. Both lines were fitted by the method of least squares.

methods other than multiple-column gas-liquid chromatography were not attempted, and so residue identifications must be regarded as presumptive.

Concentrations in the twenty samples analysed ranged from 13 to 79 parts per 10^9. These are the lowest concentrations recorded in recent analyses of marine fish; most values range from about 0·2 p.p.m. upward[2]. A summary of all analyses of *T. mexicanus* is shown in Table 1. A large number of fish were taken at station 85, thus allowing a size class analysis of fish from that location. The data from thi s station (Fig. 1) reveal an increase in total residue and DDE concentrations with body weight.

The presence of DDT residues in fish cannot be explained in the same way as it has been for mammals and birds[5], for fish probably do not produce water soluble metabolites. In studies of fish in which analytical techniques have allowed the detection of p,p'-DDMU, p,p'-DDMS, and p,p'-DDNU (metabolic precursors of the water soluble metabolites p,p'-DDA and DBP), these compounds have not been found. These precursors of normal excretory products in mammals are found in measurable quantities in known metabolizers. Also, DDT residues can enter fish through the integument or gill surfaces directly from the water[9], as well as with ingested food. Once DDT residues are in the tissues of a fish, they can pass out into the water again[10]. The relative importance of these processes of direct uptake from water, assimilation from stomach contents, and diffusive loss to water, is poorly understood. Some workers have assigned a high importance to the diffusive processes across the gill and assumed that residue contents are determined by an equilibrium between inward and outward diffusion[10,11]. Recent evidence, however, indicates that food intake is ten times more important than inward diffusion from water in determining concentrations of residues (unpublished results of K. J. Macek and S. Korn). An important finding from their work is that fish progressively accumulate ^{14}C-DDT from water without any equilibrium point being reached, which implies that diffusive loss of residues is of little consequence. If cumulative assimilation with relatively little diffusive loss is taking place, larger fish would have higher concentrations than smaller, younger fish. If there is a diffusion equilibrium between body residues and residues in water, fish would have about the same concentrations, regardless of size (provided that the ambient water contained a fixed, homogeneous concentration of residues). Our results indicate that *T. mexicanus* conforms to this cumulative assimilation model, for concentrations increase with the size of the fish.

This work was supported by a grant from the US National Science Foundation.

[1] Butler, P. A., in *The Biological Impact of Pesticides in the Environment* (in the press).

[2] Risebrough, R. W., in *Chemical Fallout, First Rochester Conference on Toxicity*, 5–23 (Thomas, Springfield, 1969); Bailey, T. E., and Hanmen, J. R., *J. Sanit. Eng. Div. Amer. Soc. Civil Engs.*, **93**, 27 (1967); Butler, P. A., *Bioscience*, **19**, 889 (1969).

[3] Weaver, L., Gunnerson, C. G., Breidenbach, A. W., and Lichtenberg, J. J., *Publ. Health Reps.*, **80**, 481 (1965); Keith, J. O., and Hunt, E. G., *Trans. Thirty-first North. Amer. Wildlife Nat. Res. Conf., 1966*, 150–177.

[4] Duffy, J. R., and O'Donnell, D., *J. Fish. Res. Board Canada*, **25**, 189 (1968); Jensen, S., Johnels, A. G., Olsson, M., and Otterlind, G., *Nature*, **224**, 247 (1969).

[5] Robinson, J., Richardson, A., Crabtree, A. N., Coulson, J. C., and Potts, G. R., *Nature*, **214**, 1307 (1967).

[6] Keith, J. O., in *The Biological Impact of Pesticides in the Environment* (in the press).

[7] Stanley, R. L., and Le Favoure, H. T., *J. Assoc. Off. Agric. Chem.*, **48**, 666 (1965).

[8] *Pesticide Analytical Manual, II* (US Dept. Health, Education and Welfare, Food and Drug Administration, revised 1968).

[9] Premdas, F. H., and Anderson, J. M., *J. Fish Res. Board Canada*, **20**, 827 (1963).

[10] Gakstatter, J. H., and Weiss, C. M., *Trans. Amer. Fish. Soc.*, **96**, 301 (1967).

[11] Holden, A. V., *J. Appl. Ecol.*, **3**, Suppl., 45 (1967).

In Vitro Conversion of DDT to DDD by the Intestinal Microflora of the Northern Anchovy, Engraulis mordax

THE tissues of many species of fish contain DDT and two of its derivatives, DDE and DDD. Fish which feed on plankton, such as the northern anchovy *Engraulis mordax*, acquire DDT residues primarily through the assimilation of food[1-3]. The reductive dechlorination of DDT to DDE and DDD has been demonstrated in three species of salmonid fishes: Atlantic salmon[4,5], cutthroat trout[6], and rainbow trout[7]. The purpose of this investigation was to determine whether the intestinal microflora of the northern anchovy is capable of dechlorinating DDT and to elucidate the relative importance of bacteria and fungi in this process.

The intestinal contents of twenty-five adult anchovies were drained into a sterile test tube to which 15 ml. of sterile Difco nutrient broth was added. Aliquots (2 ml.) were injected into twenty sterile test tubes (24 ml. capacity) equipped with vaccine caps, and the cultures divided into four experimental groups: (i) series *A* was left unaltered; (ii) series *B* was inoculated with 100 units each of penicillin and streptomycin to suppress bacterial activity; (iii) series *C* was inoculated with 100 units of mycostatin to suppress fungal activity; and (iv) series *D* was inoculated with 100 units each of penicillin, streptomycin and mycostatin. One control was run with each series, in which nutrient broth was substituted for intestinal contents.

Each culture was inoculated with 0·05 ml. of 111 p.p.m. ^{14}C-DDT ethanol solution and incubated anaerobically in a nitrogen atmosphere at 15° C in the dark for 6, 12, 24, 48 and 72 h. After digestion with a glacial acetic-perchloric acid mixture[8], 4 ml. of water was added and the DDT residues extracted with 10 ml. of hexane and concentrated to 0·5 ml. in a Kuderna Danish evaporator. The DDT residues in 0·05 ml. of the concentrate were separated by descending chromatography using 2,2,4-trimethylpentane as the mobile solvent and 8 per cent 2-phenoxyethanol in anhydrous ethyl ether as the immobile solvent[9]. The chromatograms were cut into 1 cm horizontal strips and counted with a Nuclear Chicago Unilux II scintillation counter. The peaks were identified by parallel chromatography using 1°/$_{oo}$ DDT, DDE and

Table 1. RELATIVE AMOUNT (PER CENT) OF ACTIVITY RECOVERED IN EACH FRACTION AND THE TOTAL AMOUNT (d.p.m.) OF ACTIVITY RECOVERED FROM EACH CULTURE INCUBATED WITHOUT PENICILLIN, STREPTOMYCIN, AND MYCOSTATIN (GROUP A), WITH PENICILLIN AND STREPTOMYCIN (GROUP B), WITH MYCOSTATIN (GROUP C), AND WITH PENICILLIN, STREPTOMYCIN AND MYCOSTATIN (GROUP D)

Experimental group	Incubation time (h)	Per cent Hexane fraction			Polar fraction	Particulate fraction	Total activity (d.p.m.)
		DDT	DDD	DDE			
A-1	6	45·0	36·0	0	0·9	18·1	588,825
2	12	27·3	32·7	0	1·7	38·3	593,140
3	24	30·6	35·0	0	4·0	30·4	606,768
4	48	29·5	27·9	> 0·1	2·6	40·0	605,312
5	72	25·0	54·1	> 0·1	2·2	18·7	597,120
Control	72	93·9	0	0	1·0	5·1	610,694
B-1	6	52·7	27·2	0	0·7	19·4	599,228
2	12	25·5	29·0	0	1·9	43·6	580,897
3	24	49·8	22·9	0	2·3	25·0	596,178
4	48	40·9	24·8	> 0·1	2·8	31·5	601,580
5	72	41·6	37·8	> 0·1	1·9	18·7	598,244
Control	72	98·0	0	0	0·3	1·7	618,226
C-1	6	18·1	27·6	0	1·5	52·8	597,821
2	12	23·7	35·3	0	1·5	39·5	590,254
3	24	37·0	22·1	0	1·2	39·7	609,078
4	48	34·5	38·8	> 0·1	1·6	25·1	599,890
5	72	21·3	42·7	> 0·1	2·4	33·6	596,330
Control	72	81·9	0	0	1·5	16·6	596,714
D-1	6	56·2	25·8	0	0·6	17·4	613,946
2	12	76·5	16·9	0	0·9	5·7	609,300
3	24	67·6	13·6	0	0·7	18·1	592,666
4	48	48·1	34·4	0	1·5	16·0	592,985
5	72	39·2	41·2	0	1·6	18·0	597,110
Control	72	99·4	0	0	0·2	0·4	621,109

DDD standards and $AgNO_3$ as the chromagenic agent[9].

After the extraction was completed, each culture was filtered through HA 'Millipore' filters, and the filter (particulate fraction) and filtrate (polar fraction) counted with the scintillation counter.

Most activity was recovered in the hexane fraction (Table 1). No ^{14}C-DDD or DDE appeared in the controls, and the activity found in the polar and particulate fractions was considerably less than in the test cultures. All the test cultures contained high concentrations of ^{14}C-DDD, but little or no ^{14}C-DDE. The highest levels of labelled DDD occurred in group A in which neither bacterial nor fungal activity was suppressed, and the lowest in group D where both were suppressed. Both fungi and bacteria converted DDT to DDD, and the efficacy of this conversion is shown by the high level of ^{14}C-DDD observed after only 6 h of incubation in group A. This rapid metabolism of DDT to DDD may have important consequences for survival because DDD is less toxic than DDT[10,11], and the degradation and assimilation of ingested food require several days[12].

A relatively large amount of labelled material remained in the cultures after the hexane extraction. From 18 to 53 per cent of the recovered activity was particle bound in groups A, B and C. Particle bound activity was only

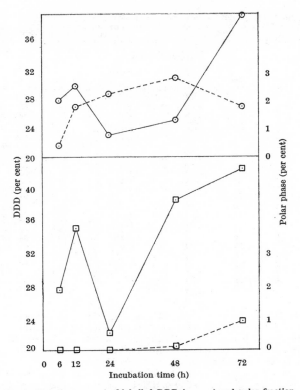

Fig. 1. Relative amount of labelled DDD (———) and polar fraction activity (- - -) recovered as a function of incubation time. Values were calculated after subtracting controls. Bacterial activity in group B (○) and fungal activity in group C (□) were inhibited.

6 to 18 per cent in group *D* where microbial activity was most inhibited. Polar phase activity was also lowest in group *D* where it constituted 0·6 to 1·6 per cent of the recovered activity as opposed to 0·7 to 4·0 per cent in groups *A* and *B*. This may indicate further metabolism of DDD to water soluble products such as DDA and the incorporation of labelled material into refractory organic compounds.

Some differential activity of the intestinal fungi and bacteria is suggested. In group *B*, where bacterial activity was suppressed, DDD levels were lower and polar phase activity higher than in group *C* where fungal activity was suppressed (Fig. 1). This indicates that while both bacteria and fungi metabolized DDT to DDD, the fungi were primarily responsible for further degrading DDD to a water soluble product in anaerobic conditions.

[1] Pillmore, R. E., *US Wildl. Res. Lab.* (1961).
[2] Robinson, J., Richardson, A., Crabtree, A. N., Coulson, J. C., and Potts, G. R., *Nature,* **214**, 1307 (1967).
[3] Woodwell, G. M., Wurster, C. F., and Isaacson, P. A., *Science,* **156**, 821 (1967).
[4] Premdas, F. H., and Anderson, J. M., *J. Fish. Res. Bd. Canada,* **20**, 827 (1963).
[5] Greer, G. L., and Paim, U., *J. Fish. Res. Bd. Canada,* **25**, 2321 (1968).
[6] Allison, D., Kallman, B. J., Cope, O. B., and van Valin, C. C., *Science,* **142**, 958 (1963).
[7] Wedemeyer, G., *Life Sci.,* **7**, 219 (1968).
[8] Stanley, R. L., and LeFavoure, H. T., *J. Ass. Off. Agric. Chem.,* **48**, 666 (1965).
[9] Mitchel, L. C., *J. Ass. Off. Agric. Chem.,* **40**, 294 (1957).
[10] Metcalf, R. L., *Organic Insecticides* (Interscience Publishers, New York, 1955).
[11] Kenaga, E. E., *Bull. Entomol. Soc. Amer.,* **12**, 117 (1966).
[12] Barrington, E. J. W., in *The Physiology of Fishes,* **1** (Academic Press, Inc., New York, 1957).

DDT: Disrupted Osmoregulatory Events in the Intestine of the Eel Anguilla rostrata Adapted to Seawater

Ralph H. Janicki

William B. Kinter

In view of the importance of marine fisheries to human nutrition (1), it is particularly alarming that fish are the most sensitive of all vertebrates to widespread organochlorine pollutants such as DDT (2). Although it is generally accepted that in both vertebrates and invertebrates DDT exerts a direct toxic effect on the nervous system, the biochemical basis is still uncertain (3). Because some evidence exists for inhibition of (Na+ and K+) activated transport adenosine triphosphatase in several organs including brain (4), we wondered if the extreme sensitivity of fish to DDT might not reflect parallel disruptions of osmoregulation and nerve function.

Marine teleosts face desiccation in their hypertonic environment. In part, they preserve tissue hypotonicity by drinking seawater and absorbing sodium and chloride across the intestinal epithelium. Water follows the absorption of these ions and is retained in the body while the ions are secreted by the gill epithelium (5). The adenosine triphosphatases appear to function in these osmoregulatory processes. The (Na+ and K+) activated, Mg^{2+}-dependent adenosine triphosphatase [(Na+,K+,Mg^{2+}) adenosine triphosphatase] which is sensitive to ouabain is believed to be involved in the transport of sodium across cell membranes (6). Supporting this hypothesis is the fact that the activity of this enzyme in the intestinal mucosa of eels adapted to seawater is twice that seen in eels adapted to freshwater (7). Furthermore, there is evidence that the mitochondrial portion of the (Mg^{2+}) adenosine triphosphatase which is the portion stimulated by 2,4-dinitrophenol is involved in oxidative phosphorylation (8). Thus by supplying ATP, this enzyme may be at least indirectly involved in active transport.

Eels (*Anguilla rostrata*) between 30

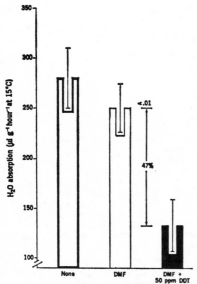

Fig. 1. The effect of DDT (final concentration 50 ppm, or 1.4 × 10⁻⁴ mole/liter) on water absorption in the intestine of the eel. The addition of 0.5 percent *N,N*-dimethylformamide (DMF) to the Ringer solution has no significant effect on intestinal water absoption ($P > .3$). The addition of 50 ppm of DDT in DMF decreases water absorption 47 percent ($P < .01$). The tops of the bars represent means; the vertical lines represent standard errors. Seven intestinal sacs were assayed in each medium.

and 38 cm long were captured in estuaries along the Maine coast. They were adapted to and maintained in seawater (13° to 16°C) for 3 weeks before use. After decapitation, the intestines from the pyloric sphincter to the anus were excised. Intestines were cannulated and prepared for measurement of water absorption by means of a procedure in which noneverted sacs are used (9). Analytical grade *p,p'*-DDT (10) was dissolved in *N,N*-dimethylformamide (DMF) at a concentration of 10 mg/ml. Three media were used: Ringer, Ringer containing 0.5 percent DMF, and Ringer containing 0.5 percent DMF and a suspension of 50 parts of DDT per million (ppm). Each cannulated intestine was first incubated in its respective medium for 60 minutes at 2° to 5°C; the medium was repeatedly rinsed through the lumen to allow DDT to enter the tissue. Then, each intestine was filled with its respective medium under slight hydrostatic pressure and sealed. The sacs were then incubated at 15°C in erlenmeyer flasks gassed with oxygen and containing additional medium. Thus, when present, DDT was on both the serosal and mucosal sides. Water absorption was calculated as sac weight at time 0 minus weight at 60 minutes and presented as microliters of H_2O per gram of intestine per hour.

When isolated sacs of eel intestine were incubated in $1.4 \times 10^{-4}M$ DDT (50 ppm), there was a 47 percent ($P < .01$) inhibition of water absorption (Fig. 1). This is a strong inhibition and is comparable to the effects of more familiar blocking agents. For example, $1 \times 10^{-4}M$ 2,4-dinitrophenol and $1 \times 10^{-4}M$ ouabain also inhibit approximately 50 percent of the intestinal water absorption in eels adapted to seawater (11). However, DDT and, to a lesser extent, 2,4-dinitrophenol are lipophilic, and since membranes contain lipids, the concentrations of these compounds at specific transport loci may have been greater.

We also correlated the impairment of water absorption with the inhibition of activities of adenosine triphosphatase in homogenates of intestinal mucosa. For each determination of (Na^+, K^+,Mg^{2+}) and (Mg^{2+}) adenosine triphosphatases the intestinal mucosae of six eels were pooled and homogenized (2 percent, weight per volume) in a mixture of $0.25M$ sucrose, $0.005M$ disodium ethylenediaminetetraacetic acid, and $0.03M$ imidazole (adjusted to pH 7.4). Deoxycholate was omitted because it partially inhibited (Na^+, K^+,Mg^{2+}) adenosine triphosphatase. The assay medium (7) contained in final concentration 20 mM imidazole (pH 7.8) and either 100 mM NaCl and 20 mM KCl, or 120 mM NaCl. For the enzyme studies, DDT was dissolved in DMF. At a final concentration of 5 percent, DMF had no effect on (Na^+,K^+,Mg^{2+}) adenosine triphosphatase and inhibited only 20 percent of (Mg^{2+}) adenosine triphosphatase activity. Before addition of ATP, the tubes were incubated for 30 minutes at 15°C (12). Reactions were initiated by the addition of 0.3 ml of a solution containing in final concentration 100 mM MgCl$_2$ and 100 mM disodium salt of adenosine triphosphate (neutralized with saturated tris buffer). The final volume was 5.0 ml. After 30 minutes at 15°C, the reactions were terminated by addition of 1.0 ml of ice-cold 30 percent trichloroacetic acid; the tubes were placed in ice for 10 minutes. The precipitate was removed by centrifugation. Because DDT interfered with the colorimetric procedure for determination of phosphate (13), it was extracted from a 3.0-ml portion of the supernatant with an equal volume of ice-cold toluene.

The DDT had a strong inhibitory effect on (Na^+,K^+,Mg^{2+}) adenosine triphosphatase (Fig. 2). Even at the low concentration of 5 ppm of DDT (1.4×10^{-5} mole/liter), there was 43 percent inhibition. The concentration of DDT which inhibited 50 percent of

the activity was approximately 15 ppm (4.0×10^{-5} mole/liter). We were unable to determine whether (Na^+,K^+, Mg^{2+}) adenosine triphosphatase can be completely inhibited by DDT. Suspensions of 250 ppm with 5 percent DMF were already very cloudy; thus it seemed unreasonable to go to higher concentrations. The (Mg^{2+}) adenosine triphosphatase in intestinal homogenates of the eel was also inhibited by DDT, albeit to a lesser extent (Fig. 2). The break in the dose response curve sug-

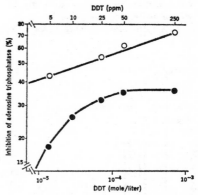

Fig. 2. The relationship between the concentration of DDT and the percentage of inhibition of (Na^+,K^+,Mg^{2+}) and (Mg^{2+}) adenosine triphosphatases in the intestinal mucosa of the eel. The log percentage of the inhibition is plotted against the log concentration of DDT in molarity and parts per million. This procedure establishes a linear relationship between activity and the concentration of the inhibitor (20).

gests the presence of two (Mg^{2+}) adenosine triphosphatases—one smaller component which is sensitive to DDT and a larger fraction which is resistant. The question arises as to whether previous environmental exposure to DDT was, to some degree, already suppressing adenosine triphosphatase activity in the eels used in the experiment. Although we do not know the extent of previous contamination, whole-body concentrations of DDT in other teleosts inhabiting the Frenchman Bay area have been reported as 0.033 ppm or lower (14), an amount which presumably would have had little effect on our data.

Intestinal water absorption in eels adapted to seawater is inhibited in vitro by cyanide, 2,4-dinitrophenol, and ouabain (11), and represents, in all likelihood, a coupled process involving sodium, the major cation transported across the intestine (15). Our data demonstrate that DDT impairs water absorption apparently by inhibiting mucosal adenosine triphosphatases involved in sodium transport. A reasonable correlation between enzyme activity and function exists, since at 15°C the normal temperature for these eels, the activity of (Na^+, K^+, Mg^{2+}) adenosine triphosphatase and water absorption are each about halved by 15 and 50 ppm of DDT, respectively. Although adenosine triphosphatases and sodium secretion in the gills may also prove sensitive, our findings have significance in that many species of teleosts are reported to die at whole-body concentrations commonly averaging between 5 and 10 ppm of DDT (16). Accordingly, the extreme sensitivity of fish to DDT may involve inhibition of adenosine triphosphatases and an accompanying disruption in osmoregulation.

The inhibition of adenosine triphosphatase activity in teleosts seems to be a general property of organochlorine insecticides. In preliminary surveys (17) DDT inhibited (Na^+, K^+, Mg^{2+}) adenosine triphosphatase, but not (Mg^{2+}) adenosine triphosphatase in the intestinal mucosae of several other marine species, and both of these enzymes in the gill of winter flounder (Pseudopleuronectes americanus). In the lake trout, the organochlorine insecticides chlordane, dicofol, lindane, and DDT inhibit (Mg^{2+}) adenosine triphosphatase in brain, liver, and muscle, and (Na^+, K^+, Mg^{2+}) adenosine triphosphatase in brain (18). Furthermore, disrupted osmoregulation, evidenced by an alteration in the concentration of serum electrolytes, has been reported in both northern puffers and goldfish treated with the organochlorine insecticide endrin (19). These observations tend to substantiate our hypothesis that the sensitivity of teleosts to organochlorine insecticides involves impairment of osmotic regulation.

References and Notes

1. J. H. Ryther, Science 166, 72 (1969).
2. R. D. O'Brien, Insecticides: Action and Metabolism (Academic Press, New York, 1967), pp. 302–304; D. W. Johnson, Trans. Am. Fish. Soc. 97, 398 (1968); C. F. Wurster, Biol. Conserv. 1, 123 (1969).
3. R. D. O'Brien, Insecticides: Action and Metabolism (Academic Press, New York, 1967), pp. 108–132; F. Matsumura, Science 169, 1343 (1970).
4. F. Matsumura and K. C. Patil, Science 166, 121 (1969); R. B. Koch, J. Neurochem. 16, 269 (1969); T. Akera, T. M. Brody, N. Leeling, Biochem. Pharmacol. 20, 471 (1971).
5. G. Parry, Biol. Rev. 41, 392 (1966); F. P. Conte, in Fish Physiology, W. S. Hoar and D. J. Randall, Ed. (Academic Press, New York, 1969), pp. 214–292.
6. J. C. Skou, Physiol. Rev. 45, 596 (1965).
7. L. M. Jampol and F. H. Epstein, Am. J. Physiol. 218, 607 (1970).
8. M. E. Pullman, H. S. Penefsky, A. Datta, E. Racker, J. Biol. Chem. 235, 3322 (1960); H. S. Penefsky, M. E. Pullman, A. Datta, E. Racker, ibid., p. 3330.
9. M. Oide and S. Utida, Mar. Biol. 1, 102 (1968).
10. Better than 99 percent pure 1,1,1-trichloro-2,2-bis(p-chlorophenyl)ethane.
11. M. Oide, Annot. Zool. Japan. 40, 130 (1967); W. C. Mackay, Bull. Mt. Desert Isl. Biol. Lab. 9, 23 (1969); ——— and R. H. Janicki, ibid. 10, 42 (1970).
12. The initial incubation period was included to allow time for DDT to permeate the fragments in the homogenate. Such incubation, for periods up to 24 hours at 5°C, did not change the degree of inhibition by DDT.
13. C. H. Fiske and Y. SubbaRow, J. Biol. Chem. 66, 375 (1925).
14. R. H. Adamson, J. B. Sullivan, D. P. Rall, Bull. Mt. Desert Isl. Biol. Lab. 8, 2 (1969).
15. E. Skadhauge, J. Physiol. 204, 135 (1969).
16. D. Allison, B. J. Kallman, O. B. Cope, C. C. Van Valin, Science 142, 958 (1963); K. Warner and O. C. Fenderson, J. Wildl. Manage. 26, 86 (1962).
17. R. H. Janicki and W. B. Kinter, Am. Zool. 10, 540 (1970); Fed. Proc. 30, 673 (1971); Nature, in press.
18. R. B. Koch, Chem.-Biol. Interactions 1, 199 (1969).
19. R. Eisler and P. H. Edmunds, Trans. Am. Fish. Soc. 95, 153 (1966); B. F. Grant and P. M. Mehrle, J. Fish. Res. Bd. Canada 27, 2225 (1970).
20. C. I. Bliss, Science 79, 38 (1935).
21. Supported by PHS grant AM 06479.

ACTION OF DDT ON EVOKED AND SPONTANEOUS ACTIVITY FROM THE RAINBOW TROUT LATERAL LINE NERVE

THOMAS G. BAHR and ROBERT C. BALL

Institute of Water Research, Michigan State University, East Lansing, Michigan
48823

INTRODUCTION

THE TREMORS, convulsions and hyperexcitability associated with DDT-poisoning reflect a neurotoxic mechanism common to many animals. A site of action of DDT appears to be on the axon membrane (Narahashi & Haas, 1968), but the relative sensitivity of sensory, central and motor neurons to DDT and their contribution to the "DDT-syndrome" is somewhat unclear. Motor fibers of DDT-poisoned cockroaches have been shown to elicit repetitive discharge following single stimuli (Yeager & Munson, 1945), but sensory fibers behave similarly at much lower DDT concentrations (Roeder & Weiant, 1946). It was hypothesized that increased motor activity in poisoned cockroaches resulted from intense afferent bombardment of motor neurons by impulses originating in sensory fibers.

The striking similarity of DDT symptoms between fish and invertebrates prompted us to investigate the effects of DDT on the nervous system of trout. In preliminary experiments we found that DDT was rapidly incorporated into trout lateral line nerves following intravenous injection and also after exposing live fish to DDT in water (Bahr, 1968).

We decided to explore the possibility of using multifiber recordings of spontaneous activity from the lateral line nerve as an index of sublethal damage to the nervous system. We hypothesized that the frequency of neural discharge from this nerve trunk might provide a sensitive physiological parameter reflecting subtle changes in the integrity of the nerve and associated receptor structures.

Although tremors and convulsions normally reflect events originating in the central nervous system (CNS) it is possible that similar symptoms could arise from post synaptic events initiated by intense afferent activity impinging on central neurons. We wanted to test this possibility by monitoring the frequency of spontaneous discharge and the patterns of evoked discharge from the lateral line nerve before and after the onset of DDT-induced tremoring.

MATERIALS AND METHODS

Rainbow trout, *Salmo gairdneri*, weighing from 186 to 288 g were anesthetized in MS 222 (100 mg/l), and permanently immobilized by severing the spinal cord at the level of the foramen magnum. They were placed on their sides in a water-filled chamber and the gills were irrigated with aerated fresh water. The water level was raised to completely cover the fish. Fish had been acclimated to a temperature of 13°C for at least 6 weeks before testing and all experiments were conducted at this temperature. Bipolar silver wire electrodes were looped under the distal end of the centrally-cut lateral line nerve at a point midway along the length of the trunk. Water was excluded from the recording site by housing electrodes in an oil-filled nerve chamber firmly secured to the skin. Neural activity was monophasically displayed on a cathode-ray oscilloscope (Tektronix 502A) after amplification (Grass P-8).

The frequency of spontaneous neural discharge was determined by electrically counting the compound action potentials on a decade scaler. The rise time of potentials was reduced with a bistable multivibrator enabling the signals to pass the input gate of the scaler. Frequency measurements by this method agreed very well with those determined from photographs of oscilloscope displays. Counting efficiency was increased by randomly eliminating a number of active fibers by partially cutting through the nerve a few millimeters distal to recording electrodes or by completely severing the nerve a few centimetres distal to the recording electrodes. These procedures reduced the level of normal spontaneous activity and prevented much of the overlapping of potentials which otherwise would have occurred at the recording site. Because this procedure may have biased our recordings by eliminating potentially more DDT-sensitive fibers, we periodically compared results obtained from pared nerves with results from intact-nerve preparations. We found no evidence to indicate that pared nerve preparations were biasing our results.

Evoked responses were generated by directing drops of water to the surface of the water overlying the innervated portion of the lateral line. Each water drop electrically triggered the sweep of the oscilloscope beam allowing every evoked neural response to appear at the same position along each sweep of the beam. Evoked responses were recorded by superimposing 40 low-intensity displays on film.

DDT was administered to fish in water and by injection into the blood. The DDT was a 99-plus per cent pure p,p′ isomer of 1,1,1-trichloro-2,2-bis (p-chlorophenyl)ethane, M.W. 306.48, obtained from the City Chemical Corporation of New York. Individual fish were placed in 5-gallon aquaria containing 15 ppm DDT. DDT was added in an ethanol solvent and was well mixed before fish were added. They were removed after 6 hr and lateral line nerve activity was examined. Eight fish were used, four serving as controls.

In separate experiments DDT was introduced directly into the blood. This was accomplished by first mixing a DDT–ethanol solution with Courtlands saline (Wolf, 1963), then immediately injecting 0·4 ml of the resulting suspension into the posterior cardinal vein through a cannula inserted near the tail. The ethanol concentration was 10 per cent in all injections. Doses used were 0·01, 0·1, 0·2, 0·4, 0·5, 1·0 and 2·0 mg DDT per injection. Two fish were tested at each level (each weighing approximately 200 g) and six additional fish received control injections of saline (10% ethanol). Assuming a blood volume of 2·25 per cent of the live weight of the fish (Schiffman & Fromm, 1959) the initial concentration of DDT in the blood was calculated to range from 2 to 400 ppm. This method of administering DDT proved to be very good despite the fact that crystals of DDT were present in the injected fluid. No abnormalities could be detected in electrocardiograms after DDT injection indicating that the heart did not experience much stress from injections. Oils proved to be poor carriers for DDT; their high viscosity rapidly induced cardiovascular failure.

RESULTS

Aquaria experiments

Trout exposed to DDT in water quickly demonstrated the neurotoxic nature of the insecticide. After 1 hr fish became hyperexcitable and demonstrated periods of violent unco-ordinated swimming movements with very apparent erratic eye and jaw movements. These symptoms increased in severity for about three hours although later became less pronounced. By the end of the 6-hr exposure period fish were noticeably calmer but still demonstrated erratic eye movements, opercular trembling and occasional trembling in the trunk.

Electrophysiological examination of lateral line nerve preparations revealed no apparent differences between treated and untreated fish. The duration and amplitude of the evoked responses appeared to be normal in all of the fish tested and the frequency of spontaneous activity appeared normal.

Injection experiments

Results of these experiments were similar to the aquaria experiments. The frequency of spontaneous neural discharge showed no increase or decrease throughout a 2-hr post injection period (Fig. 1). In similar plots of discharge frequency

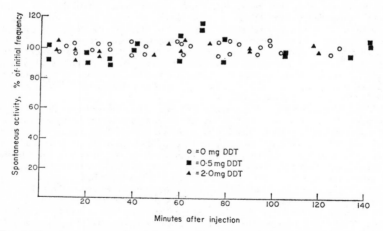

FIG. 1. Spontaneous neural activity from the rainbow trout lateral line nerve after intravenous injections of DDT.

176

against time, for each injection level and for controls, the slope of the least squares regression line was not significantly different from zero ($0.05 < P$). Evoked neural activity from the lateral line nerve of DDT-injected fish was also unchanged (Fig. 2).

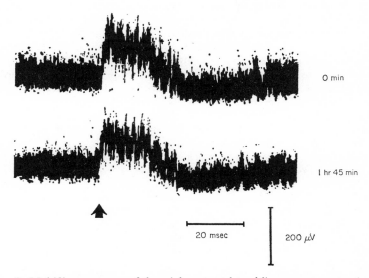

O min

1 hr 45 min

20 msec

200 μV

FIG. 2. Multifiber responses of the rainbow trout lateral line nerve to water drop stimuli (arrow). Each recording represents 40 superimposed oscilloscope traces. Top: evoked response before DDT injection. Bottom: evoked response 1 hr 45 min after intravenous injection of 2 mg DDT. At this time tremoring was evident.

Although we could detect no abnormalities in lateral line discharge after DDT we did observe tremoring in fish receiving injections of 0·5 mg DDT or more. Tremoring was restricted to the head region above the level of spinal transection and was characterized by spasmodic twitching of the eyes and operculum. Tremoring began approximately 1 hr after injection. The symptoms were presumably caused by DDT because they were absent in controls and in fish receiving smaller doses of the chemical.

DISCUSSION

Our findings were somewhat surprising and appear to conflict with results of a similar study conducted by Anderson (1968). Anderson demonstrated a marked prolongation of evoked lateral line activity after brook trout were exposed to DDT. This study, however, is not entirely comparable in that our fish were acclimated to a temperature 8°C cooler. The possibility exists that acclimation temperature could affect the differential uptake of DDT into various portions of the nervous system such that uptake into the lateral line nerve might be favored at higher temperatures. However, a more likely explanation for the lack of apparent response in our fish concerns the magnitude of the lateral line involvement under the thermal regime of our experiments. Anderson found that the prolongation of lateral line evoked responses was increasingly greater when the temperature was reduced from the acclimation temperature. Testing the lateral line nerve at temperatures near the acclimation temperature would result in a decrease of this prolongation. In our experiments we analyzed the evoked lateral line response, as well as spontaneous activity, at the same temperature to which our fish had been acclimated (13°C). According to Anderson's data we should not have expected much prolongation of the evoked response and our data reflect this. Very slight differences in duration of evoked bursts of activity are difficult to measure because of uncertainty in determining the point corresponding to the end of the evoked burst. This can be seen by examining Fig. 2.

It might be argued that DDT was not present in adequate amounts in the nervous system of our fish or that enough time was not allowed for its neurotoxic action to manifest. Our observations are not consistent with this premise. Had the lateral line nerve been an important peripheral component of the DDT-poisoning syndrome we should have observed electrophysiological changes in the nerve during, if not before, the onset of tremoring. Since we observed obvious tremoring in the absence of lateral line involvement, we believe that the lateral line nerve did not play an important role in the classical motor symptoms of acute DDT-poisoning. Using the frequency of spontaneous neural discharge from multifiber lateral line nerve preparations as an index of DDT damage to the nervous system does not appear to be a very sensitive technique to assess the neurotoxic effects of DDT. There is evidence to indicate that the CNS is the primary target for DDT in fish (Anderson & Peterson, 1969) and our results are consistent with these findings.

Acknowledgements—This research was sponsored by a Predoctoral Fellowship (FI-WP-26, 004-03) from the Federal Water Pollution Control Administration and represents part of a Ph.D. thesis submitted to Michigan State University, Department of Fisheries and Wildlife, by the senior author.

REFERENCES

ANDERSON J. M. (1968) Effect of sublethal DDT on the lateral line of brook trout, *Salvelinus fontinalis*. *J. Fish. Res. Bd. Canada* **25**, 2677–2682.

ANDERSON J. M. & PETERSON M. R. (1969) DDT: Sublethal effects on brook trout nervous system. *Science* **164**, 440–441.

BAHR T. G. (1968) Electrophysiological responses of the lateral line and heart to stresses of hypoxia, cyanide, and DDT in rainbow trout. Ph.D. thesis, Michigan State University, East Lansing.

NARAHASHI T. & HAAS H. G. (1968) Interaction of DDT with components of lobster nerve membrane conductance. *J. gen. Physiol.* **51**, 177–198.

ROEDER K. D. & WEIANT E. A. (1946) The site of action of DDT in the cockroach. *Science* **103**, 304–306.

SCHIFFMAN R. H. & FROMM P. O. (1959) Measurement of some physiological parameters in rainbow trout (*Salmo gairdnerii*). *Can. J. Zool.* **37**, 25–32.

WOLF K. (1963) Physiological salines for fresh water teleosts. *Prog. Fish-Cult.* **25**, 135–140.

YEAGER J. F. & MUNSON S. C. (1945) Physiological evidence of a site of action of DDT in an insect. *Science* **102**, 305–307.

Key Word Index—Lateral line nerve; rainbow trout; insecticides; DDT; spontaneous neural discharge; evoked neural discharge.

179

DDT Effects in Invertebrate Systems

Laboratory and Field Evaluation of the Persistence of Some Insecticides on Noctuid[1] Larvae on Apple in Norway

Torgeir Edland

For the past 25 years DDT has been widely used in controlling insect pests injurious to fruit trees. Because it is a highly persistent insecticide, a single application is often sufficient to control many pests that occur during a long period, whereas less persistent insecticides may require several treatments to obtain the same efficiency. However, recent years have brought evidences of DDT having detrimental effects on certain wild animal species. Hence, as an attempt to reduce the great potential hazard from such environmental contamination, any use of DDT after 1 Oct. 1970 in agriculture and horticulture in Norway was prohibited by law.

To find an alternative to DDT, several experiments on control of actual pests have been carried out in Norway during the past few years. The experiments reported here, conducted in 1969, were designed to evaluate the persistence of some organo-

---

[1] Lepidoptera: Noctuidae.

phosphates in comparison with DDT against certain lepidopterous pests on apple.

MATERIALS AND METHODS.—*Insects.*—Ca. 3000 3rd- and 4th-stage larvae of the noctuids *Eupsilia transversa* Hufnagel, *Orthosia gothica* L., and *Xylina vetusta* Hübner, which all are common pests on fruit trees in Norway, were used as test insects. The larvae, which originated from eggs produced by light-trap-caught females, were reared on apple leaves in the laboratory prior to the tests.

Insecticides.—Commercial formulations of azinphosmethyl (25% WP), bromophos (36% EC), DDT (50% WP), diazinon (20% EC) fenthion (50% EC), malathion (60% EC), and parathion (35% EC) were used. All concentrations, which are listed in Table 1, refer to the amounts of active ingredient in the dilute spray and were those normally recommended in the lowest rate for field use. An exception was made for bromophos, which was used in a concentration of 0.036% AI in the 1st tests, but was reduced to 0.018% AI for the spray applied 25 June.

Laboratory Tests.—Apple leaves, which received insecticides on 16 June in field plots, were fed to larvae in petri dishes at 20–25°C. Two to 5 larvae of the same species were enclosed in each dish, with 4 replications for each species and insecticide. The 1st trial was started 3 hr after spraying, followed by trials 1, 2, 3, and 5 days later. The leaves were collected from the sprayed trees on the day in which they were fed to the larvae. Thus, the insecticidal residues grew progressively older during a given test. Larval mortalities were recorded daily. In these experiments 580 larvae were tested.

Field Screening Tests.—Young apple trees were sprayed to runoff on 4, 16, and 25 June. Each insecticide was applied to 4 replicated 5-tree plots. At different intervals after the applications, larvae were confined in nylon net bags on treated and untreated branches on all replicates. The bags, enclosing 2–5 larvae of each species, were mostly examined every 2–3 days for assessing mortalities. Climatic data for the testing periods were obtained from the weather station at Ås, situated 350 m from the experimental orchard.

RESULTS.—Neither the laboratory tests nor the field tests showed any significant differences between the species tested. Therefore, the following results are based on the sum of data obtained for all 3 species.

Laboratory Tests.—Fig. 1A–E shows graphically the results from these tests. Larval mortality, expressed as percentages and corrected by Abbott's (1925) formula for the mortality of the untreated controls in each test, is plotted against time after larval exposure to the spray residues.

When the larvae were exposed to the residues within 2 days after spraying, 80–100% were killed by azinphosmethyl, DDT, bromophos, and parathion, while mortality caused by malathion, diazinon.

Treatment	% AI	Trial and age (days) of residues of initial exposure of larvae		% larval mortality on days after exposure to residues		
		Trial	age	2	5	8
DDT	0.1	A	1	100	100	
			3	100	—	
			5	—	100	
			9	100	100	
			12		96	93
		B	(3 hr)	88	100	100
			1	83	85	100
			2	94	100	88
			4	70	85	60
			6	69	62	63
Azinphosmethyl	.037	A	1	58	—	
			3	93	100	
			5	—	100	
			9	62	100	
			12	—	98	100
		B	(3 hr)	81	88	90
			1	96	100	91
			2	100	81	62
			4	73	62	38
			6	65	—	81
Bromophos	.036	A	1	66	56	
			3	98	71	
			5	—	88	
			9	62	25	
			12	—	50	83
	.018	B	(3 hr)	25	18	94
			1	46	—	40
			2	44		50
			4	5		28
Parathion	.014	A	1	18		
			3	13		
			12	30	65	33
		B	(3 hr)	17	50	65
			1			54

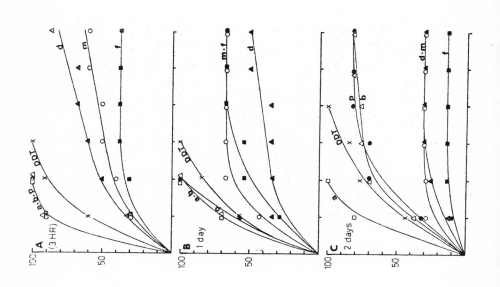

LARVAL MORTALITY (PER CENT.)

184

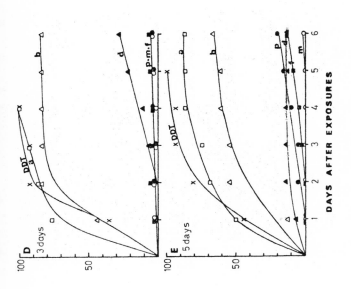

Table 1.—Effects of different spray residues on noctuid larvae in field trials, 1969. Trial A was sprayed 4 June, trial B 25 June, and the larvae were exposed to the residues at different intervals.

Compound	Rate	Trial	Interval			
Diazinon	.02		2	0	0	0
			4	25	17	8
		B	(3 hr)	19	13	0
			1	13	4	0
			2	31	25	19
Malathion	.072	B	(3 hr)	60	50	40
			1	50	42	29
			2	13	13	0
Fenthion	.05	B	(3 hr)	45	25	10
			1	54	50	17
			2	50	50	31
Untreated		A	1	38	24	10
			3	18	11	4
			5	0	0	—
			9	0	0	0
			12	36	—	—
		B	(3 hr)	5	5	5
			1	13	8	0
			2	6	6	6
			4	32	26	10
			6	14	11	5

Fig. 1.—Effects of 7 insecticides on noctuid larvae exposed to the residues at 3 hr, 1, 2, 3, and 5 days after application. Best-fit lines are drawn by eye. ☐ a, azinphosmethyl; △ b, bromophos; ● p, parathion; ▲ d, diazinon; ■ f, fenthion; ○ m, malathion; × DDT.

and fenthion was considerably lower. In fact, the 3-hr-old residues of the 3 last-mentioned chemicals were less active than those of DDT and azinphos-methyl at an age of 5 days. DDT was the most persistent insecticide, closely followed by azinphos-methyl. Bromophos, too, showed high persistence compared with parathion, diazinon, malathion, and fenthion, which had practically lost their larvicidal activity after 3 days from application.

Field Tests.—Table 1 summarizes the main results from these tests. In Trial A, where the trees were sprayed 4 June, DDT caused 100% mortality within 2 days on larvae exposed to the residue no later than 9 days after application. Twelve days or more after spray application, the residue of DDT still maintained a high larval toxicity. Azinphosmethyl and bromophos showed very similar and high residual effects. However, both chemicals acted significantly more slowly than DDT, and showed a considerably shorter residual effect. The larvicidal activity of parathion was rather poor, causing a final maximum mortality of 50%. The residual effects of the remaining insecticides were similar to those of parathion or even lower, and their results have therefore been deleted from the table.

The maximum temperatures on the 1st and 2nd day after spraying in Trial A (5 and 6 June) were only 15.5 and 16.9°C. Then it increased considerably, varying between 23.7 and 26.9°C during the following 10 days. No rainfall occurred during this testing period.

The results obtained in Trial B corresponded well with those in Trial A. During this testing period, the daily maximum temperatures were high, 25.8°C on 25 June when the sprays were applied and above 20°C for the rest of the period, except for a temperature of 17.5°C on 28 June. Only very light rainfalls were recorded during the actual period, with a maximum of 1.8 mm on 1 day (29 June).

Under these conditions there were no differences in the larvicidal activity of DDT and azinphosmethyl. Neither caused 100% mortality within 2 days, but all larvae recorded as living at that time exhibited clear symptoms of being strongly poisoned. For bromophos, where the dilute spray concentration was only ½ that used in previous tests, the larval mortality was somewhat lower than for the other persistent insecticides. Parathion and the remaining insecticides showed poor larvicidal activities, even where the larvae were exposed to the residues only 3 hr after treatment. At later exposures only insignificant differences were found between those low persistence organophosphates and the untreated controls.

DISCUSSION.—On basis of the larval mortality caused by residues of different ages, the 7 insecticides tested in these experiments fell into 2 groups. In the 1st

186

group, DDT, azinphosmethyl, and bromophos showed a rapid, high, and long-lasting toxicity compared with parathion, diazinon, malathion, and fenthion in the 2nd group.

Residues of DDT seemed to be most toxic and killed the larvae more rapidly at low temperatures than did those of the organophosphates. This is well in line with earlier reports (Guthrie 1950, Maier-Bode 1965).

In persistence, azinphosmethyl and bromophos were superior to the other organophosphates. However, both the initial toxicity and the residual effects of bromophos were considerably reduced when the dilute spray was applied at the lower concentration.

All the chemicals of the 2nd group showed very short residual effects. Possibly because of the high temperatures, the residues seemed to be greatly decomposed even 3 hr after application. Consequently the larval mortality was quickly reduced and was often insignificantly different from that of the untreated controls only 2–3 days after spraying.

It should be emphasized that the present results concern residual effects upon half-grown noctuid larvae only, which may be considerably more resistant to spray residues than are young lepidopterous larvae and other pests. In other field experiments (Edland, unpublished data) all the insecticides used in the present study gave satisfactory control of young noctuid and geometrid larvae when applied on trees already infested. Such direct insecticidal effects, however, do not indicate any residual effect.

Several pests may infest the fruit trees a short time after spraying. Evidences of heavy reinfestation of newly sprayed orchards by airborne larvae of the winter moth, *Operophtera brumata* (L.), have been obtained in Western Norway (Edland 1971). As such cases normally involve young larvae, it is probable that low persistent insecticides may cause higher mortalities for a longer period than those obtained on the relatively large noctuid larvae in the present experiment. However the differences in relative persistences should be the same.

From this study it may be concluded that none of the insecticides in the 2nd group are persistent enough to give satisfactory control of pests attacking the plants some time after spraying. This statement is in agreement with earlier findings. For instance, in experiments on control of the apple fruit moth, *Argyresthia conjugella* Zeller, which attacks its host plants from the middle of June to the end of July, parathion, diazinon, and malathion gave no apparent control from single sprays carried out in July, whereas azinphosmethyl gave 90–100% control and was more effective than DDT (Edland 1965).

Both azinphosmethyl and bromophos are probably well suited to replace DDT for chemical control of many chewing pests, and they are likely to be used as excellent alternatives also to other chlorinated hydrocarbons.

REFERENCES CITED

Abbott, W. S. 1925. A method of computing the effectiveness of an insecticide. J. Econ. Entomol. 18: 265–7.

Edland, T. 1965. Apple fruit moth (*Argyresthia conjugella* Zell.). Biology and control—A preliminary report. [Reprint with English summary.] Gartneryrket 55: 430–6.

1971. Wind dispersal of the winter moth larvae *Operophtera brumata* L. (Lep., Geometridae) and its relevance to control measures. Norsk Entomol. Tidsskr. 18: 103–5.

Guthrie, F. E. 1950. Effect of temperature on toxicity of certain organic insecticides. J. Econ. Entomol. 43: 559–60.

Maier-Bode, H. 1965. Pflanzenschutzmittel-Rückstände. Insecticide. Verlag Eugen Ulmer, Stuttgart. 456 p.

AUTHOR INDEX

KEY-WORD TITLE INDEX

RA recent articles &
research RIP
in progress

Guide to Current Research

The research summaries appearing in the following section were obtained through a search of the Smithsonian Science Information Exchange data base conducted in June, 1973.

The Exchange annually registers 85,000 to 100,000 notices of current research projects covering a wide range of disciplines and sources of support. SSIE endeavors to retain up to two full years of current research information in its active file. The selection of summaries appearing in this section does not represent the complete SSIE collection of information on this topic, but, rather, has been specifically tailored to reflect the scientific content of this particular volume. A limited number of summaries may have been omitted because clearance for publication by the supporting agency or organization was not received prior to the publication date.

SSIE is the only, single source for information on ongoing and recently terminated research in all areas of the life, physical, behavioral, social and engineering sciences. The SSIE file is updated daily by a professional staff of scientists utilizing a comprehensive and flexible system of hierarchical indexing. Retrieval of subject information is conducted by these same specialists, using computer-connected, video display terminals which allow instant access to the entire data base and on-line refinement of search strategies. SSIE offers an information service unequalled anywhere: comprehensive and vital information on who is conducting what research where and under whose support.

More current information, and in some cases expanded coverage, on the topic considered in this volume is available directly from SSIE. This information is offered at modest cost in the form of custom searches of the SSIE file designed specifically to meet the user's need or as an update of the subject search in this section. For more information on SSIE, contact MSS or write directly to the Smithsonian Science Information Exchange, 1730 M Street, N.W., Washington, D.C. 20036. Subject search or updated package requirements may be discussed with SSIE scientists by calling the Exchange at (202) 381-5511.

THE ROLE OF MICROFLORA IN THE PERSISTENCE
AND DECOMPOSITION OF PESTICIDE RESIDUES,
M. ALEXANDER, State University of New York,
Agricultural Experiment Sta., Ithaca, New
York 14850

OBJECTIVE: Identify the intermediates
and products formed during the microbial
decomposition of pesticides. Investigate
the mechanisms of microbial degradation of
pesticides.

APPROACH: Herbicide and insecticide
degrading microorganisms will be isolated
from soil. The ability of these isolates to
degrade the pesticide will be examined by
use of intact cells and enzyme preparations.
Cells or enzymes will also be incubated
with suspected or proven intermediates to
determine the subsequent fate of the
chemicals. The products of the reactions
will be identified by paper, thin-layer and
gas chromatography, and by IR, UV, NMR and
mass spectrometry. Cometabolism as a means
of herbicide and insecticide transformation
in soil will also be investigated.

PROGRESS: DDT is converted by
microorganisms in vitro to 4,4'-
dichlorobenzophenone and other products
anaerobically, and one ring is cleaved
microbiologically in the presence of O(2) to
yield p- chlorophenylacetate acid. An
Arthrobacter was isolated which is able to
extensively degrade p-chlorophenylacetate
and cleave its ring; in the process, a
metabolite identified as
p-chlorophenylglycolaldehyde is produced.
No p-chlorophenylacetate accumulation is
likely in nature owing to its rapid
decomposition by microorganisms. Means of
exploiting cometabolism to enhance DDT
biodegradation and the effect of structural
analogues on the decomposition have been
established. In the metabolism of 2,4-D,
the first chlorine atom of the herbicide is

removed by a lactonizing dehalogenase to form a chlorobutenolide. The second chlorine seems to be removed from a chlorosuccinate intermediate. In the decomposition of 2,4-D in natural environments, 2,4-dichlorophenol and an as yet unidentified persistent compound are formed, while a trichlorophenol is produced from 2,4,5-T. The di- and trichlorophenols are not persistent. Several analogues of benzonitrile herbicides are subject to microbial metabolism.

NAVY ENVIRONMENT: MICROBIAL DEGRADATION OF DDT,
M. ALEXANDER, State University of New York, School of Agriculture, Ithaca, New York 14850

THE DEPARTMENT OF THE NAVY IS RESPONSIBLE FOR ALL PESTICIDES IT USES. THIS RESPONSIBILITY INCLUDES PESTICIDE WASTE PRODUCTS WHICH FIND THEIR WAY INTO THE MARINE ENVIRONMENT. THIS EFFORT IS DIRECTED TOWARD FINDING MEANS OF ENHANCING NATURALLY OCCURRING BIOLOGICAL DEGRADATION.

THE PRINCIPAL INVESTIGATOR WILL: (1) DETERMINE MORE COMPLETELY THE PRODUCTS FORMED AS DDT IS METABOLIZED MICROBIOLOGICALLY, (2) DETERMINE WHICH OF THE PRODUCTS ARE FORMED BY THE MICROBIAL COMMUNITIES OF SEA WATER, (3) ASSESS THE EFFECT OF ENVIRONMENTAL CONDITIONS ON THE RATE OF ATTACK ON DDT BY MARINE MICROORGANISMS AND (4) ATTEMPT TO ESTABLISH THE REASON FOR THE LONG PERSISTENCE OF DDT IN NATURE.

SUPPORTING AGENCY ADDRESS INFORMATION: OFFICE OF NAVAL RESEARCH 443 ARLINGTON, VA. 22217

FATE OF PESTICIDAL POLLUTANTS, WITH
REFERENCE TO MICROBIAL DEGRADATION, UNDER
OCEANIC CONDITIONS,
G.M. BOUSH, Univ. of Wisconsin, Agricultural
Experiment Sta., Madison, Wisconsin

OBJECTIVE: Determine the effects of
certain persistent chlorinated hydrocarbon
pesticides (DDT and analogs and dieldrin and
isomers) on oceanic ecosystems. We will be
especially concerned with both pollutants
and effects on microorganisms in the surface
film ecosystem.
APPROACH: We will first determine
present levels of toxicants and related
compounds in selected marine organisms from
four study areas in Hawaii. We will then
attempt to determine the role of marine
plankton in the accumulation and/or
degradation of these toxicants, as well as
the effects these compounds may have on
specific marine plankton components.
PROGRESS: To study the transformation
process of stable chlorinated hydrocarbon
pesticides in marine environments, samples
of seawater, bottom sediments from both
ocean and estuarines, surface films, algae,
and marine plankton were collected and
treated with radiolabeled insecticides at
the collection sites and incubated for 30
days in the laboratory. Also various
microorganisms were isolated from these
samples and their metabolic activities on
the insecticides were monitored along with
some laboratory cultures of unicellular
algae. Transformation of DDT and cyclodiene
insecticides took place in samples with
biological materials such as surface films,
plankton, and algae, but not in waters from
open ocean. A number of marine
microorganisms in pure culture also
metabolized the pesticides. The patterns of
metabolic activities by the microorganisms
were similar to those observed in the

194

field-collected samples. Anacystis
nidulans, a freshwater blue-green alga, has
been found to tolerate sodium chloride (1
percent by weight) and DDT
1,1,1-trichloro-2,2-bis-(p-chlorophenyl)
ethane (800 parts per billion) separately,
but growth was inhibited in the presence of
both compounds. This inhibition was
reversed by an increased calcium
concentration. It is possible that
inhibition of (Na ion, K ion)- activated
adenosine triphosphatase) by DDT causes this
species to lose the ability to tolerate
sodium chloride.

EFFECTS OF PESTICIDES ON NON-TARGET
ORGANISMS,
J.W. BUTCHER, Michigan State University,
Agricultural Experiment Sta., East Lansing,
Michigan 48823

OBJECTIVE: Assess the effect of
insectide applications on selected species
of microarthropods which are important in
the process of organic decomposition.
Examine the effect of insecticides on
selected species of microarthropods under
laboratory conditions in regard to their
biology, development, and ecological
displacement. Investigate the effect of
insecticides on the interaction between
phytophagous insect species and the plant
parasite of jack pine, Melampryum lineare.
APPROACH: Monitor Collembola, Acari,
and Opiliones species within dieldrin
treated and untreated woodlots in Monroe
County. Determine changes in abundance and
species complex in relation to residue
persistence. A series of test plots have
been established on South campus to study
the effects of dieldrin and DDT on the
abundance and changes in the species complex

of Collembola, Acari, and other soil arthropods. These plots are on cultivated land and will be planted to corn yearly. Collembola and Acari are being cultured under laboratory conditions. Individual species are being used to determine their role in the process of organic decomposition. The consequence of insecticide application on this process is being studied. Studies are underway to determine the effect of feeding of various phytophagous insect species on Melampryum lineare. Field plots have been set up and differences in feeding due to insecticide application are being compiled.

PROGRESS: Studies are continuing on biology and systematics of soil arthropods and their role in pesticide degradation. Recent work at Michigan State University has indicated that soil-inhabiting Collembola and Acarina may play significant roles in DDT degradation. Of the various groups of microarthropods collected, the Collembola and the oribatid mites were found to show the most consistent pathways of DDT metabolism and movement. In all of the releases, microarthropods showed only traces or barely measurable levels of p,p'-DDE by the end of the sampling periods. The results of this work indicate that both major isomers and the major metabolite of DDT are degraded, or at least not accumulated, by forest litter and soil microarthropods. The time required at the very low levels used here may be as short as four to five weeks. Apparent metabolic pathways of this degradation in both macro- and microarthropods under field conditions are also shown.

REDUCTION OR ELIMINATION IN COMMERCIAL
CHANNELS OF ADVERSE EFFECTS OF PESTICIDE
RESIDUES ON FOOD,
J.T. CARDWELL, Mississippi St. University,
Agricultural Experiment Sta., State College,
Mississippi 39762

OBJECTIVE: Not Provided
APPROACH: Not Provided
PROGRESS: Repeated trials to
demonstrate symbiotic or associative action
of certain bacterial strains especially
Streptococous lactis and Lactobacillus
acidophilus on the dechlorination of DDT
below DDD failed to produce consistent
results. The inconsistencies were probably
more closely associated with improper
analytical techniques than to effect of
treatments. It was found that fermented
dairy products could be manufactured from
DDT contaminated milk although there was an
inhibition of acid development attributable
to DDT. The mechanism of such inhibition
awaits further investigation.

PESTICIDE RESIDUE EFFECTS ON LARVAL MARINE
FISHES,
C.F. COLE, Univ. of Massachusetts, School of
Agriculture, Amherst, Massachusetts 01002

Effects of chronic, low-level pesticide
exposures on eggs and larvae of the winter
flounder (Pseudopleuronectes americanus,
Walbaum) are being evaluated. A previous
segment of this project examined chlorinated
hydrocarbon residue levels in winter
flounder from the Weweantic River estuary, a
small tidal tributary of Buzzards Bay,

Massachusetts. The effects of these residue
levels on the reproductive success of the
flounder using this estuary as spawning and
nursery grounds is of particular concern.
Field investigations have demonstrated that
certain insecticides (DDT and dieldrin)
commonly used in the watershed by cranberry
culturing and mosquito control programs are
present in the flounder; heptachlor is also
present but its source is unknown.
Increasing concentrations of these
insecticides were noted in flounder ovaries
during the months prior to spawning.
Laboratory investigations are presently
underway to determine what concentration
ranges of the insecticides residues in
gonadal tissues generate measurably adverse
effects during reproduction. The survival
of eggs, larvae, and juveniles spawned from
dosed adults is being followed and the
reproductive success is then to be evaluated
in light of concentrations from gonads of
similarly treated adults. Gas-liquid
chromatography is the major analytical tool.
  The flounder dosing and rearing experiments
are being undertaken at the Bureau of Sport
Fish and Wildlife Laboraory, Narragansett,
Rhode Island.

MODE OF ACTION OF INSECTICIDES AS RELATED TO
INSECT METABOLISM AND BEHAVIOR,
L.K. CUTKOMP, Univ. of Minnesota,
Agricultural Experiment Sta., Saint Paul,
Minnesota 55101

    OBJECTIVE: Determine the effects of
insecticides, particularly chlorinated
hydrocarbons and pyrethrum on the enzyme
systems, adenosine triphosphatase (ATPase)

both in vitro and in vivo. The insecticidal
effects and the ATPase enzyme activity will
also be related to periodic (rhythmic)
physiological events in the insect.

APPROACH: Accepted current biochemical
determinations will be made of the ATPase
enzyme system in nervous and muscle tissues
of American cockroaches, honey bees and
other insects. Known dilute concentrations
of DDT, other chlorinated hydrocarbons and
pyrethrum will be tested for specific
inhibitory effects on the enzyme system. In
addition, the specific activity of the
enzyme system and insecticide sensitivity
will be related to periodicity in the
American cockroach, using established light
and dark lighting regimes during a 24-hour
day.

PROGRESS: In vitro studies of the
ATPase enzyme system in nerve cords, brains
and muscles of insects show that DDT and
related compounds produce maximum inhibition
to mitochondrial Mg ATPase (oligomycin-
sensitive) of muscle; about 3X more DDT is
required in nervous tissue. In contrast,
DDT-injected roaches (in vivo) showed about
a 33% increase of Mg ATPase while
immobilized (stressed) roaches showed
increases of 200% of mitochondrial Mg
ATPase from muscles and 50% of Na -K ATPase
from nerve cords. The mitochondrial Mg
ATPase sensitivity to DDT was doubled in
immobilized roaches. DDT tested in vitro
was more effective on the sensitive muscle
mitochondrial Mg ATPase at 27 than at 17
or 37 C. The same enzyme from nerve cords
was about 30% more sensitive to DDT at 37
than at 27 . Summarizing, DDT-inhibitory
differences found in vitro as influenced by
temperature do not correspond to negative
temperature effects found from DDT toxicity
to insects. Biorhythmicity studies with
flour beetles, Tribollum confusum, show a
peak (acrophase) of oxygen consumption near
midnight which is 6 hours after the

beginning of the dark period, using a 12:12
light-dark regimen. The beetles are also
more sensitive to a rapid-acting
insecticide, dichlorvos, about midnight.
Field-collected mosquitoes, Aedes vexans,
were slightly more tolerant to some
insecticides than in previous years.

INFLUENCE OF CHEMICAL PESTICIDES ON FOREST
COMMUNITIES,
D.L. DAHLSTEN, Univ. of California,
Agricultural Experiment Sta., Berkeley,
California 94720

OBJECTIVE: Ascertain the effect of
pesticides, particularly the sub-lethal
effects, on the various developmental stages
of birds (i.e., adult, eggs, nestling,
fledgling and juvenal); discover the links
in the movement of pesticides through food
chains in selected forest environments;
investigate the impact of pesticides on
non-target arthropods, specifically the
induced imbalance in parasite-host and
predator-prey relationships; determine
changes in floral diversity, if any, and the
causes of change.
APPROACH: Determine effects of
chemical pesticides on stability of forest
communities. Important components of the
study are determination of residue levels in
birds, movement of toxins through food
chains and impact on parasitoid-host
relationships. Ten permanent study areas
permit research on an "area-basis" rather
than "species-basis". Some areas have
history of chemical treatment. Nest boxes

200

for chickadees are used to compare birds
populations. Aphid-parasitoid fluctuations
are being studied. Light traps are used in
treated and untreated areas to prepare an
index of stability based on number of
species and number of individuals in each
species of nocturnal Lepidoptera. One
untreated area will be treated in three
years in an attempt to induce instability.
    PROGRESS: Studies one year following
the spray application were continued at
Alturas. Four projects are currently
underway. A MS listing the occurrence and
abundance of the 324 species of nocturnal
macrolepidoptera from 4 years of trapping
has been completed. The diversity index is
near completion using these species. The
geometrids were the most severely affected
in the DDT area and may be the best
indicators of disruption. Soil
microarthropods were sampled systematically
in treated, untreated, and burned-over
areas. Additional plots were established by
applying chemicals with a back-pack sprayer
at high application rates to check for gross
changes. Effects of DDT on forest pond
organisms were studied. Recovery rates of
these animals were recorded. Dispersal of
notonectids was studied in detail. In the
avian studies high nest box acceptance rates
were recorded and nestling mortality was
highest (47%) in the DDT-treated plot.
Scale populations were declining in another
study at South Lake Tahoe due to natural
enemies.

INFLUENCE OF PESTICIDAL CHLORINATED
HYDROCARBONS UPON MECHANISM IN POULTRY,
M. DELAK, Univ. of Zagreb, Zagreb,
Yugoslavia

    OBJECTIVE: Determine the effect of

chlorinated hydrocarbons on the adaptation mechanism in poultry.

APPROACH: The intensity of reaction of the adrenal cortex to various chlorinated hydrocarbons will be determined. The amount of residue in tissue of treated animals will also be studied. Later the reaction of the adaption mechanism to specific amounts of pesticide, to various duration of application will be determined. Also different chlorinated hydrocarbons and combinations of them will be studied.

PROGRESS: DDT or its metabolites accumulate to a significant extent in the adrenals of fowl. Even when no additional DDT was added to feed (i.e., only amounts in feed from insecticide usage), almost 2 ppm of total DDT metabolites were found to accumulate in the gland. At lower levels of feeding, the predominant metabolite in the gland is p,p'-DDE, which is a second step breakdown product of DDT. However, at higher levels of feeding, there appear to be greater proportions of p,p'-DDT which may indicate that capacity for metabolism of DDT by the adrenal has been exceeded at higher level dosages, even though it continues to accumulate the insecticide. Total DDT and metabolite concentrating capacity in ppm of tissue, compared to other tissues when birds were fed 50 ppm in the diet were: (1) Adrenal vs brain, 20:1; (2) Adrenal vs liver 2:1. In view of past results which showed that o,p'- DDT and o,p'-DDD depress adrenal steroid synthesis, it is possible that such accumulations could reduce normal responses to stressful environments.

ECOLOGY OF ANIMALS IN DDT-CONTAMINATED ENVIRONMENTS,
J.B. DIMOND, Univ. of Maine, School of Agriculture, Orono, Maine 04473

Research will be continued on ecological effects on natural animal populations of chronic contact with low levels of DDT residues. The residues derive from large scale applications made for control of forest insects several years ago. In spite of the passage of time, residues still persist.

Species selected for study include trout, robins, the red-backed salamander, and a crayfish. Comparisons of population structure are being made between populations in sprayed and unsprayed plots. Several other parameters are being studied with particular emphasis on reproductive phenomena.

IMPACT OF PESTICIDES UPON SEMI-WILD ECOSYSTEMS,
J.B. DIMOND, Univ. of Maine, Agricultural Experiment Sta., Orono, Maine 04473

OBJECTIVE: Continue studies of the ecological significance of long-lasting DDT residues to contaminated forest animals; monitor the effects on the forest ecosystem of fenitrothion applied for control of the spruce budworm.

APPROACH: As a result of DDT applications for spruce budworm control in several previous years, areas can be found with widely ranging levels of long-lasting DDT contamination in soils and several organisms. Population success of animal species in heavily contaminated areas will be compared to lightly contaminated areas. Special attention will be paid to reproductive effects. Fenitrothion has replaced DDT as the recommended chemical for spruce budworm control. Little is known of the environmental impact of this new

chemical. Several phases will be studied
including effect on aquatic insect
populations, on parasites of the budworm,
and the effectiveness of the chemical itself
in controlling the target insect.

PROGRESS: Work on the effects of
chronic exposure to low levels of DDT to
several forest dwelling animals continued.
The approach involves studying population
dynamics of species in DDT-contaminated
areas compared to untreated areas with
minimal contamination. Species involved are
a crayfish, Cambarus bartoni, the red-backed
salamander, robins, and shrews. Work in
1971 involved establishment of sampling
plans and sampling to determine population
age structure. Special studies were devoted
to reproductive processes. None of these
have reached completion. This work is
partially supported by EPA. Where Zectran
was applied to plots to evaluate control
potential against the spruce budworm,
studies were initiated on the impact on
budworm parasites, on predators and
associated invertebrates in the fir crown
community, and on aquatic insects. This
work was supported by the U.S. Forest
Service. Effects of Zectran on aquatic
insects are relatively mild, compared to
DDT. Studies of effects on parasites and
predators will be continued for another
year.

EFFECT OF LONG TERM SUB-LETHAL DOSES OF
PESTICIDES ON BIOCHEMISTRY OF MARINE
ORGANISMS,
R.H. ENGEL, William F. Clapp Laboratories,
Duxbury, Massachusetts 02323

The purpose of this study is to examine
the effects of long term sub-lethal doses of
pesticides on enzyme systems, pathology,

respiration, and reproductive cycles of the quahog, M. mercenaria. At the present time, the pesticides being used are lindane and DDT.

The study was started in January 1969, and will run for an undetermined length of time. Funding is on an annual basis.

PESTICIDE EFFECTS ON THE BEHAVIOR AND ECOLOGY OF HERMIT CRABS,
W.C. GRANT, Williams College, Graduate School, Williamstown, Massachusetts 01267

Two species of hermit crabs, Pagurus acadianus and P. pubescens are abundant on rocky bottoms in Frenchman Bay, Maine where they have become locally sympatric in recent years. As large detritus feeders they are important elements of the sub-littoral communities of rocky shores. Indication that such hardy species were threatened by pesticide contamination would be a matter of grave concern.

The investigation will have two related objectives: 1) to determine effects of sub-lethal doses of organochloride pesticides (DDT) on hermit crab behavior. Hermit crabs have highly stereotyped patterns of behavior related to shell selection, aggressions, etc. such that neural trauma caused by external agents should be readily detectable. 2) To study several aspects of niche diversification in P. acadianus and P. pubescens using data from both laboratory and field investigations. These will include studies of shell selection, shell availability, conspecific and interspecific competition for shells and comparative homeostatic ability (measured by endogenous respiratory rate of tissue homogenates).

EFFECTS OF POLYCHLORINATED BIPHENYLS AND DDT
ON PELICANS,
Y.A. GREICHUS, South Dakota State University,
School of Arts, Brookings, South Dakota
57006

    Young pelicans will be taken from
nesting sites and administered 100 mg/day of
polychlorinated biphenyls or 50 mg/day of a
combination of DDT, DDD and DDE in the same
proportions as those found in fish taken
near nesting sites. During the last two
weeks of the study, the birds will be
stressed by reducing their ration by
one-half. During the study the learning
ability of the birds will be assessed by
conditioning to various symbols and
rewarding for proper response.
    Blood samples and fat and muscle
biopsies will be taken during the study and
at the end. Bibds will be sacrificed after
eleven weeks, examined for ecto-and
endo-parasites and gross and microscopic
pathology. Levels of polychlorinated
biphenyls and DDT, DDD and DDE will be
determined in various tissues and in whole
bodies of treated and untreated pelicans.
Polychlorinated biphenyl and insecticide
residues, blood chemistry values and
parasites of wild pelicans will be compared
to those of the penned birds. Relationships
between residue levels, number and types of
parasites, pathologic conditions of tissues,
blood chemistry levels and behavior will be
examined and analyzed statistically.

PHYSIOLOGICAL AND BEHAVIORAL EFFECTS OF DDT,
DDD AND DDE ON PENNED CORMORANTS,
Y.A. GREICHUS, South Dakota State University,
Agricultural Experiment Sta., Brookings,
South Dakota 57006

OBJECTIVE: Determine the levels and
tissue distribution of organochlorine
pesticides in various tissues of penned
double-crested cormorants (Phalacrocorax a.
auritus) fed these pesticides and relate
these to the pathological and clinical
findings and to the individual and social
behavior of these birds. Also, determine
the effects of ingested pesticides on the
parasitic fauna of the birds.

APPROACH: Nestling cormorants taken
from a natural rookery will be caged and fed
fish to which have been added graded levels
of an insecticide mixture. Insecticide
level, clinical, pathological and behavioral
studies will be made on the birds in an
effort to determine what treatments effects
there are. In addition, parasite counts
will be made to determine whether ingestion
of the mixture at the levels chosen has an
effect.

PROGRESS: During the summers of 1969
and 1970, forty-two nestling cormorants were
placed in cages on the South Dakota State
University campus and three levels of a
combination of DDE,DDD and DDT were
administered to the treatment groups.
During the study, weights and blood samples
were taken regularly. At the close of the
study, birds were necropsied and tissue
samples taken for pathologic examination and
determination of insecticide levels.
Internal and external parasites were
collected from all birds. Blood
chemistries, determination of levels of
insecticides in the whole body, brain and
feathers and the numbers and types of
parasites for all birds in the 1969 study

207

have been completed and data is being statistically analyzed. Analysis of the information gained from the 1970 study will not begin until all data from the 1969 study has been processed.

TOXICITY OF POLLUTANTS TO STRIPED BASS, J.S. HUGHES, State Div. of Fish & Game, Baton Rouge, Louisiana 70804

Overall Study Objective: To study possible causes of mortality of striped bass eggs, fry and fingerlings and determine means of preventing mortality.

This Job Objective: (1) To determine the toxicity of pollutants to various sizes of striped bass.

Procedures: Striped bass ranging in size from fry to six inches in length will be used for these toxicity studies. Standard bioassay techniques will be followed. Dilution water will be from Bayou DeSiard, a lake adjacent to the laboratory. The water quality will be checked before each series of tests. Bioassays will be run at 22 degrees Centigrade.

Pollutants that will be included in this study are paper mill wastes; oil well effluents; wastes from sugar cane mills; rotenone; antimycin A; and pesticides, such as, 2,4-D, 2,4,5-T, silvex, endothal, diquat, dylox, methyl parathion, DDT, and endrin.

Total survival, total mortality and the TLm will be reported for 24, 48, 72, and 96 hours.

THE EFFECTS OF DDT, TOXAPHENE AND
POLYCHLORINATED BIPHENYL ON THYROID AND
REPRODUCTIVE FUNCTION IN THE BOBWHITE QUAIL
COLINUS VIRGINIANUS,
J.G. HURST, Okla. St. Univ., Graduate School,
Stillwater, Oklahoma 74075

This study will attempt to determine if
the chemicals being tested affect thyroid
and reproductive function in the bobwhite
quail. The effects on thyroid function will
be determined by comparing the weights,
histology, and 2-hour I 131 uptakes of the
thyroid of the treated quail with those of
control birds. The reproductive function
will be ascertained by determining egg
production, hatchability of the eggs, and
survival of the chicks hatched. At necropsy
adrenal weights, the testes weights of
males, and general status of the
reproductive system in females will also be
noted.

BIOLOGICAL SIGNIFICANCE OF PESTICIDE
RESIDUES IN FISH,
H.E. JOHNSON, Michigan State University,
Agricultural Experiment Sta., East Lansing,
Michigan 48823

OBJECTIVE: Determine the effect of
specific pesticides in varied combinations
on the growth and reproduction of fish and
fate and effect of pesticide residues in
fish subjected to environmental stress.
Identify the source of pesticide residues in
selected inland water of Michigan.
APPROACH: Fish will be exposed to
graded series of pesticide concentrations

under continuous flow conditions to measure
effects on gonad development, egg fertility,
and embryo survival. Stress conditions
(thermal, osmotic, and diet) will be
superimposed on exposure conditions and
effects evaluated by analysis of growth,
pathology and reproductive success. Blood
chemistry will be monitored as an indicator
of residue mobilization.

PROGRESS: Coho salmon fry from Lake
Michigan parent stock suffered significantly
higher mortality during the period from
hatching to early feeding than fry from Lake
Huron or Oregon stocks. Mortality among the
Lake Michigan fry was less when reared at 13
C than at 5 C, 9 oC or 17 oC. At colder
temperatures the development of Lake
Michigan fry was delayed to a greater extent
than fry from the other sources. DDT
residue concentrations were higher in Lake
Michigan eggs and fry than in the other
stocks. Rainbow trout (steelhead) fry
exposed to DDT at the time of hatching did
not have increased mortalities at any of the
four rearing temperatures. Toxicity
bioassays of DDT-polychlorinated biphenyl
(PCB) mixtures with coho salmon fingerling
indicates additive effects of the toxicants.
Concentrations of PCB and DDT which were
sublethal in separate tests caused
significant mortality when tested in
combination. Where DDT concentrations were
held constant the mortality increased in
direct proportion to the concentration of
PCB. Sublethal Dieldrin concentrations have
not increased the susceptibility of coho
salmon or rainbow trout to fungus infection
in laboratory tests.

BEHAVIORAL TOXICOLOGY INDUCED BY
INSECTICIDES,
H. KLEEREKOPER, Texas A & M University
System, School of Science, College Station,
Texas 77843

To determine, through a study of
locomotor patterns and orientation behavior
of fish, whether behavior control mechanisms
in vertebrates are affected by exposure to
sublethal concentrations of chlorinated
hydrocarbons and organic phosphates.

To ascertain whether exposure to
sublethal concentrations of these substances
affects sensory mechanisms and their central
integration by verifying the effects of such
exposure on the normal ability of fish to
use olfactory and acoustic stimuli in
orientation.

To relate the locomotor behavior to
concentration and exposure time to DDT and
parathion at various temperatures.

EFFECTS OF PESTICIDES ON MICROORGANISMS
IMPORTANT TO THE DAIRY INDUSTRY,
B.E. LANGLOIS, Univ. of Kentucky,
Agricultural Experiment Sta., Lexington,
Kentucky 40506

OBJECTIVE: Compare the effect of
organochlorine pesticides on the growth and
metabolic activity of microorganisms in
media with and without casein. Determine
and characterize the mode of action of
organochlorine pesticides in inhibiting
growth and metabolic activity of

microorganisms.

APPROACH: (a) Milk and broth media
will be used to determine the effects of
organochlorine pesticides on the growth and
metabolic activity of microorganisms
isolated from milk as well as those used in
the manufacture of various food products.
The microorganisms will be divided into
groups based on their resistance or
sensitivity to pesticide inhibition in order
to determine if inhibition is at the species
or genus level and if dependent on a
particular pesticide. The metabolic
activity affected as well as the fate of the
pesticide during the growth period will be
determined. (b) Sensitive microorganisms
from (a) will be used to characterize the
mode of action involved in the inhibition.
Studies will be made to determine if
inhibition is due to action of the
pesticides on some structural organization
of the cell or to reaction with participants
in enzyme systems.

PROGRESS: The effect of DDT, Dieldrin,
Endrin, chlordane and heptachlor on the
growth of 40 species of food related
bacteria in skimmilk and broth is being
determined. Preliminary results indicate
that the growth of all bacteria in skimmilk
is not affected by up to 100 mug/ml of the
pesticides. However, the growth of
micrococci, staphylococci and bacilli in
broth is affected by the pesticides.

MODIFICATION OF THE MARINE ENVIRONMENT BY
MAN,
R. LASKER, U.S. Dept. of Commerce, Southwest
Fisheries Center, San Diego, California
92037

Technical Objective: It is known from
a variety of sources that man's modification

212

of the ocean altered the environment of
resident pelagic fishes and their food.
Evidence exists that marine organisms
concentrate substances such as DDT, lead,
mercury, polychlorinated biphenyls, among
others, and the detrimental effects of these
substances, particularly in high
concentrations, is widely known and
evidenced. The purpose of this study is to
attempt to clarify the mechanisms of
concentration of these substances, to
explain the manner in which they affect
fishes and their food, and to determine the
levels of concentrations of these man-made
pollutants in the trophic levels of the sea
leading to pelagic fish.

Approach: Survey historical trends of
DDT and related chemicals in the California
Current utilizing the California Cooperative
Fisheries Investigation (CalCOFI) plankton
samples that have been collected and stored
for over two decades. Additionally,
determine the pathways (aerosols, outfalls,
ships) whereby polychlorinated biphenyls and
pesticides enter the California Current and
are accumulated by fishes; to identify which
organic compounds currently in production
may be potential pollutants and are being
accumulated by organisms living in the
California Current; and to develop
techniques for identifying and measuring
pesticides and other man-made organic
pollutants.

Progress: The historical buildup of
DDT in the ocean off southern California is
being traced through analyses of specimens
of the myctophid fish, Stenobrachius
leucopsarus, taken on CalCOFI cruises from
1950 until the present. About 500 fish have
been analyzed and this phase of the work is
about 90 percent complete. Tentative
results indicate that the DDT gradually
built up in this species over the 20- year
period. DDT concentrations are much higher
in fish taken near the point source of

213

contamination (the Los Angeles County sewer system) and decline away from the source. Also DDT is the dominant pesticide during the early fifties, but DDE (a breakdown product of DDT) becomes more and more dominant after the mid-fifties indicating a metabolic breakdown of DDT in the ecosystem of the Los Angeles Bight.

EFFECT OF PROCESSING PROCEDURE ON MILK PESTICIDE RESIDUES WITH EMPHASIS ON RESIDUE REMOVAL,
R.A. LEDFORD, State University of New York, Agricultural Experiment Sta., Ithaca, New York 14850

OBJECTIVE: Isolate representative microorganisms of the diverse flora associate with different varieties of cheese and to determine if their growth produces any degradation of chlorinated pesticides.
APPROACH: Different appropriate cheese types will be made with active isolates to ascertain pesticide degradation under commercial conditions.
PROGRESS: The incidence of DDT and DDE degrading isolates of Brevibacterium linens and Geotrichum candidum was highest in Liederkranz cheese. The pesticide degrading activity was found to be non-inducible and not sensitive to reducing agents such as glutathione. Glucose and magnesium had no stimulatory effect on the degrading activity of the isolates in broth culture. Cells of these isolates did not bind the pesticides. Pesticide degradation was observed to be most active at pH 8.5. In a previous report of cheese studies, where DDT was added to the cheese milk and measurements of the residual DDT were made during the ripening

process, casein inhibited the pesticide.
This inhibition was found to be relieved by
trypsin activity.

ACTION MECHANISMS OF INSECTICIDAL
DERIVATIVES,
F. MATSUMURA, Univ. of Wisconsin,
Agricultural Experiment Sta., Madison,
Wisconsin

OBJECTIVE: Explore the possibilities
of using insect pathogens or any other
naturally occurring compounds as
insecticides. Study the mechanisms of
action of various chlorinated insecticides
on the nervous systems of animals.
Investigate the fates of chlorinated
hydrocarbon insecticides in the environments
with particular emphasis on the microbial
degradation of various pesticides.
APPROACH: The non insecticidal
materials proposed for investigation are
hormones, insect pheromones, chemosterilants
and other naturally occurring insecticidal
analogs. The major approach here is to
conduct laboratory evaluation of new
materials. In some cases attempts will be
made to identify the toxic or active
principles involved. Study the mode of
action of insecticidal analogs, efforts will
be made first to investigate the direct
effects of biochemical means. Also efforts
will be made to study the degradation
enzymes which metabolize insecticidal
substrates in various animals.
PROGRESS: It was found that DDT
inhibits ATPases of various organisms. To
study the mechanisms of action of DDT and
related compounds in the nervous system, the

215

nerve cords were taken from the walking legs
of the American lobster. Its nerve Na , K -
ATPase, as well as other associated ATPases,
were studied. The lobster nerve cords
contain ATPases which are more sensitive to
DDT than DDE. Efforts were also made to
study the mode of action of dieldrin in the
cockroach nerve cords. Autoradiographic
studies at the electron microscopic level
were used to study the movement of the
labeled insecticide. It was concluded that
the interstrain difference in susceptibility
toward dieldrin is likely caused by the
ability of the nervous system of the
resistant strain to accumulate less amount
of dieldrin. Also efforts were made to
establish a centrifugal method to separate
various subcellular components of the
nervous system of the American and German
cockroach. In vivo toxicity of the dieldrin
and its analogs were correlated to in situ
toxicity of the cyclodiene analogs by using
electrophysiological means.

MECHANISMS OF INSECTICIDE DEGRADATION,
F. MATSUMURA, Univ. of Wisconsin,
Agricultural Experiment Sta., Madison,
Wisconsin

   OBJECTIVE: Study the fates of
persistent pesticides particularly
chlorinated hydrocarbon insecticides in
environments. The metabolic alteration will
be the No. 1 item, but, if the necessity
arises, translocation absorption and other
physical transformation such as the change
brought to the pesticides by ultraviolet ray
and food chain processing.
   APPROACH: The major approach will be
through laboratory tests on capabilities of
various organisms to alter the fates of

216

pesticides. Radioisotopic techniques will be employed to monitor the metabolic and other changes brought by the test materials. According to the test results in the laboratory a specific semi- and full scale field test will be conducted. Specific processes or phenomena will be examined, for instance some particular "terminal residues" that have been indicated to form as the result of lab. tests would be examined as to their presence in certain key ecosystems.

PROGRESS: The major emphasis of the study was placed on the metabolic fates of chlorinated hydrocarbon insecticides. Heptachlor was found to be metabolized by the rat to form first heptachlor epoxide and then to 1-exo-hydroxy-2,3-epoxy-4,6-chlordiene. This metabolite was identified as a result of NMR, mass-spectroscopic and infrared spectroscopic analyses. Phenobarbital, DDT and dieldrin were found to stimulate the activity of protein synthesis in vitro in the cultured cells derived from Aedes aegypti larvae. The induction processes did not appear to be closely related to DNA synthesis. The increase in the activity of the protein synthesis caused by these agents apparently resulted in the increase of protein contents of the individual cells, rather than the increase in the cell number per culture. To study the mechanism of action of dieldrin attempts were made to establish a centrifugal method of separate various subcellular components of the nervous system of the American and German cockroaches. Those fractions were then examined for their capacities to bind C-dieldrin. An autographic technique to facilitate electron microscopic studies on the behavior of labeled insecticides in the nerve homogenate was also developed. To study the mechanisms of action of DDT, nerve ATPases of lobster axon were examined.

MICROBIAL DEGRADATION OF PESTICIDES,
F. MATSUMURA, Univ. of Wisconsin,
Agricultural Experiment Sta., Madison,
Wisconsin

OBJECTIVE: Investigate the role of
microorganisms in the degradation of certain
of the more persistent pesticides such as
dieldrin, endrin and DDT.

APPROACH: First, to locate, evaluate
and identify microorganisms with
pesticide-degrading capabilities, and to
identify metabolites resulting from such
degradation. Secondly, to determine the
biochemical pathways and enzyme systems
involved. Finally, to select and develop
variants with higher pesticide-degrading
capabilities and investigate the utilization
of such organisms.

PROGRESS: As a result of an
examination of over 300 microbial cultures,
it was concluded that the acaricides,
Chlorobenzilate and Chloropropylate, were
metabolized by a yeast, Rhodotorula
gracilis. Metabolites were identified as
4,4'-dichlorobenzilic acid and 4,4'-
dichlorobenzophenone. The probable steps of
degradation are Chlorobenzilate or
Chloropropylate - 4,4'- dichlorobenzilic
acid - 4,4'- dichlorobenzophenone, although
some other intermediate metabolites might
exist. Chlorobenzilate was more easily
hydrolyzed than Chloropropylate, so that
larger amounts of carbon dioxide and 4,4'-
dichlorobenzophenone were obtained from
Chlorobenzilate degradation. Regardless of
acaricides used, longer incubation caused a
higher accumulation of 4,4'-
dichlorobenzophenone. It appears that the
decarboxylation of 4,4'- dichlorobenzilic
acid to 4,4'- dichlorobenzophenone was
hindered by alpha-ketoglutarate and enhanced
by succinate. Twenty microbial cultures
which had been shown to degrade dieldrin

218

were tested to determine their ability to degrade endrin, aldrin, DDT, gamma isomers of benzenehexachloride (gamma-BHC), and Baygon. All isolates were able to degrade DDT and endrin, whereas 13 degraded aldrin. However, none of them were able to degrade Baygon or gamma-BHC.

DDT RESIDUES IN EGGS AND BODY TISSUES OF HENS,
K.N. MAY, Univ. of Georgia, School of Agriculture, Athens, Georgia 30601

OBJECTIVE: Effect of forced molting and/or induced hyperthyroidism on elimination of DDT from eggs and tissues of hens contaminated with the pesticide.

APPROACH: DDT will be given to hens by capsule in amounts equivalent to .5, 1. and 5. ppm of feed intake. Some groups will be force molted by severe feed restriction and others will be fed thyroprotein. The DDT content of the tissues and eggs will be determined by gas chromotography to determine effect of treatments on elimination of residues.

PROGRESS: Hens consuming 20, 30 or 40 ppm of DDT daily feed intake for five days stored the DDT in the depot fat. DDT residues in eggs were 25 to 30% of the residue in the depot fat. Depletion of DDT residues from eggs and depot fat follows the kinetics of a first order reaction. The half life of DDT residues in depot fat and eggs were 7.27 and 7.59 weeks respectively. When hens were forced molted by starvation the half life decreased to 5.86 and 4.98 weeks. Feeding 0.45 mg. per day of thyroxine for 71 days decreased the half life of the residues in depot fat and eggs.

219

Combination of force molting and thyroxine
feeding did not accelerate DDT depletion
over that of thyroxine alone.

EFFECTS OF CHLORINATED HYDROCARBONS ON REEF
CORALS AND A CORAL PREDATOR,

L.R. MCCLOSKEY, Walla Walla College,
Graduate School, College Place, Washington
99324

The investigations supported by this
grant concern the physiological effects of
certain chlorinated hydrocarbons (p, p'-DDT,
Dieldrin and Aroclor 1254) on reef corals,
and the transfer of these compounds to a
primary coral predator "Acanthaster planci."
The objectives of this research are: (1)
to obtain data on present levels of
chlorinated hydrocarbons in reef corals and
in "Acanthaster"; (2) to document the
physiological effects of each organochlorine
and possible synergistic or additive effects
if more than one organochlorine is present;
and (3) to determine if the organochlorines
are selectively accumulated by coral
predators. The results of this research
project will enable the investigator to
determine whether alleged starfish epidemics
now plaguing Pacific coral reefs are
correlated with higher organochlorine
levels.

DEGRADATION ENZYMOLOGY OF ENVIRONMENTAL
POLLUTANTS,

L.L. MCDOWELL, State University of New York,
School of Forestry, Syracuse, New York 12224

OBJECTIVE: Find organisms that can
provide a source for enzymological studies
on the degradation of pesticides such as
DDT, DDD, parathion, etc. Establish a basis
for future assessments of the potential
effects of synthetic xenobiotics upon the
biota.

APPROACH: Organisms from soil, water
and sewage are being screened to determine
their ability to degrade or transform
various "recalcitrant" molecules.
Enrichment cultures and co-metabolism
studies are being analyzed by gas
chromatography and radiotracer techniques.

PROGRESS: Numerous species of yeasts,
filamentous fungi and bacteria representing
groups of organisms normally inhabiting
lakes, streams and soils were screened to
determine their ability to degrade or
cometabolize DDT, DDD, parathion, etc. Some
yeasts were found to absorb DDT but no
organism tested was able to degrade the
above particles.

BIOLOGICAL EFFECTS OF ENVIRONMENTAL FACTORS
IN BIRDS AND ANIMALS,

L.Z. MCFARLAND, Univ. of California,
Agricultural Experiment Sta., Davis,
California 95616

OBJECTIVE: Evaluate acute and chronic
toxicity of chlorinated hydrocarbon and
organophosphate pesticides in birds.

221

APPROACH: Study chronic effects of pesticides, newly hatched Japanese quail will receive pesticide in starter mash during a six week test period. At weekly intervals, treated and untreated birds will be removed for study. Morphological changes including weights of brain, liver, kidneys, gonads, adrenals, thyroids and pituitary will be evaluated. Biochemical changes will be studied including hexokinase, glutamic dehydrogenase, aldolase and other enzymes. Physiological measurements (EEG, heart rate, respiratory rate, etc.) will be made at 2, 4 and 6 weeks of age. Acute effects of pesticides will be studied in waterfowl. A dose will be sought that will cause death in 60 minutes when given intravenously in vegetable oil. Physiological measurements will be evaluated.

PROGRESS: Observation with the scanning electron microscope during 1971 indicated there are definite structure aberrations in eggshells of Japanese quail (Coturnix c. japonica) formed under the influence of both chlordecone (Kepone) and technical DDT. Our studies show that the influence of these pesticides may be independent of nominal thickness, and severe defects in shell structure occurs. The major defects are observed in the spongy layer and outer cuticle. Of major interest, the structural aberrations were diagnostically different between DDT and chlordecone. Currently we have begun SEM analysis of mallard (Anas platyrhynchos) eggs formed under the influence of p,p'DDE, p,p'DDT + PCM (polychlorinated biphenyls), and PCB. In conjunction with controlled experiments we have completed SEM analysis of brown pelican (Pelecanus occidentalis) eggs from three regions of North America, and have begun SEM evaluations of eggshells from three species of Falconiformes.

BIOCHEMICAL EFFECTS OF DDT AND CARBOFURAN
ANALOGS ON POND ECOSYSTEMS,
R.L. METCALF, State Natural History Survey,
Urbana, Illinois 61801

This investigation will involve the
study of new biodegradable analogs of DDT
and carbofuran in natural earthen pond
ecosystems. These selective compounds will
be radiolabeled and allowed to cycle
throughout the pond for a period of 30 days.
Samples of fish, snails, algae, water
fleas, and plankton will be seeded into the
system. Subsequently, samples of these
organisms and water will be analyzed to
compare their metabolism and ecological
magnification. The work will hope to show
how appropriate these materials might be in
larvical pest control programs and to
provide basic information on principles of
biodegradability.

ROLE OF MIXED FUNCTION OXIDASES IN
INSECTICIDE ACTION,
R.L. METCALF, Univ. of Illinois, School of
Life Sciences, Urbana, Illinois 61801

Work will be continued on the role of
the mixed function oxidases in insecticide
action with emphasis on insecticide
selectivity and biodegradability.
Asymmetrical analogues of DDT will be
evaluated for biodegradability in the model
ecosystem and in mice and insects. The
effects of additional substituent groups
such as CN, CHO, and NO2 on the DDT nucleus
will be studied. A specific effort is being
made to identify the site and type of attack
of insect and mammalian MFO enzymes on the

various weak points of the DDT molecule to
as to systematically develop further
knowledge of the principles of
biodegradability. Similar studies are being
conducted with bioisosteric carbamate
molecules including carbaryl, MOBAM, and
benzofuranyl N-methylcarbamate. The
metabolic products resulting from metabolism
alone and after treatment with a variety of
synergists are being identified.

The MFO enzymes are being
quantitatively studied in the snail,
daphnia, mosquito larvae, and fish of the
model ecosystem to provide quantitatively
for the analysis of the role of each
organism in biodegradability.

MODE OF ACTION OF INSECTICIDES -
ELECTROPHYSIOLOGICAL,
T. NARAHASHI, Duke University, School of
Medicine, Durham, North Carolina 27706

The mechanism of action of various
insecticides on the nerve and muscle system
will be studied by means of
electrophysiological methods. DDT related
compounds will be studied by means of
voltage clamp techniques to elucidate the
ionic mechanism of action and to clarify the
structure-activity relation. For this
purpose, we plan to develop a better and
more precise voltage clamp method with
crayfish giant axons utilizing internal
longitudinal axial wire. The well-known
negative temperature dependency of DDT
action will also be analyzed by that method.
The study of nicotine will be extended to
other derivatives to establish the
structure-activity relation for the effect

on the axon and end-plate. The study of
pyrethroids will be extended to elucidate
the mechanism of action on junctions, using
crustacean synapses which are highly
sensitive to them.

PHYSIOLOGY OF ESTUARINE ORGANISMS,
D.W. NIMMO, U.S. Dept. of Commerce, Natl.
Oceanic & Atmos. Admin., Gulf Breeze,
Florida 32561

The purpose of this research is to
determine the physiological effects of low
levels of pesticides on estuarine fauna. At
present, the research is divided into two
areas of interest: (1) the effects of DDT on
protein metabolism in shrimp and (2) the
kinetics and deposition of C14-labeled
pesticides in the tissues of shrimp.

MECHANISMS OF INSECTICIDAL ACTION,
R.D. OBRIEN, State University of New York,
Agricultural Experiment Sta., Ithaca, New
York 14850

OBJECTIVE: Understand the precise
mechanism by which insecticides,
particularly chlorinated hydrocarbons,
produce their toxic effects in mammals and
insects.
APPROACH: The research for anti-enzyme
activity in chlorinated hydrocarbons has
been most unsuccessful, and the bulk of
evidence suggests that these are agents
which interfere with the organizational

array in conducting membranes, rather than on enzymes. It is proposed to study the role of complex formation in the action of chlorinated hydrocarbons. Two initial procedures will be followed. One is to construct artificial membranes from materials which will provide a matrix into which one can insert membraneous material derived from mammalian and insect nerve tissue. The ion permeability and the electrical properties of these reconstructed membranes will be examined, particularly with a view to finding what properties are bestowed upon the system by the naturally derived component. Appropriate agents (including insecticides, transmitter substances, and selected drugs) will then be applied to the systems, and changes in the above properties, caused by complex formation between agent and natural membrane constituent, will be looked for.

PROGRESS: We have now studied a number of additional analogs of DDT, with respect to their ability to modify the potassium conductance which valinomycin bestows upon a synthetic (lecithindecane membrane). With one exception, the nine highly apolar compounds substantially antagonized the increased conductance with the ratios initial/final conductance varying from 1.6 for chloro DDT to 2.8 for DDT itself. The only exception was that o,p'-DDD had no effect at all upon the system. The five analogs which has polar substituents, including OH and NH(2) and NO(2) all gave substantial increases in conductance, so that the ratio initial/final conductance varied from 0.34 OH-DDT to 0.62 for NO(2)-DDT. These findings made it unlikely that the mechanism of action of DDT was what we had initially supposed, that is to say formation of a complex with valinomycin. We are therefore studying the alternative hypothesis, that these agents modify the properties of the membrane within which the valinomycin moves.

226

CHLORINATED HYDROCARBONS - PATTERNS AND
EFFECTS UPON THE REPRODUCTIVE CAPACITY OF
ANTARCTIC PELAGIC SEA BIRDS,
H.S. OLCOTT, Univ. of California, School of
Agriculture, Berkeley, California 94720

Studies approved under GA-14202 for a
first season in the Antarctic Peninsula
included plans for studies on pollution
ecology on Kerguelen Island. Because cruise
rescheduling delayed the latter plans until
early 1971, support is now required for
participation in "Eltanin" Cruise 47 and the
cooperative research on pollution ecology on
Kerguelen Island. This activity is
coordinated by Dr. Risebrough, U/California,
Berkeley, and Dr. Hubert Ceccaldi of the
Station Marine d'Endoume et Centre
d'Oceanographie, Marseilles, France, as
American-French cooperative research. In
addition, field work is planned at Cape
Hallett for comparative data in analyzing
the Palmer Station materials collected in
1970. Since polychlorinated-biphenyl (PCB)
and DDT are now most abundant in the marine
ecosystem, the Antarctic and Subantarctic
studies will make an important contribution
on global fallout patterns by studying
seabirds with restricted ranges in isolated
areas (Kerguelen) and by sampling the wide
ranging Wilson's petrel (Palmer and Hallett
Station).
Dr. Robert Risebrough, Research
Associate and an assistant will work at
Hallett Station October-December 1970, and
proceed from New Zealand to Australia to
join "Eltanin" Cruise 47 for 30 days' work
on Kerguelen Island. Travel in New Zealand
and Australia to collect local samples and
to Marseilles, France, for analyses of
Kerguelen data is approved in connection
with this research. USARP transportation
will be provided from California to New
Zealand, and New Zealand to Freemantle,
Australia.

DDT AND DIELDRIN CONCENTRATIONS IN GREAT
LAKES LAKE TROUT AND COHO SALMON,
R. REINERT, U.S. Dept. of The Interior, Bur.
of Sport Fish. & Wildlife, Ann Arbor,
Michigan 48107

The Great Lakes Fishery Laboratory has
been studying insecticide concentrations in
Great Lakes lake trout and coho salmon since
1965. DDT concentrations were studied in
different sizes of fish, different parts of
fish, and in fish from different areas of
the lake.
Both species containing high
concentrations of DDT (DDT, DDE, DDD). Adult
coho average 12-14 ppm. Twenty-two 26-inch
lake trout average 16- 19 ppm. DDT
concentrations increase as the fish increase
in size. Fish of a given size class from
the southern end of Lake Michigan contain
higher concentrations of DDT than those from
the northern end. The portions of fish
which contain the highest concentrations of
DDT are those which are highest in percent
oil. The concentrations of DDT in whole
coho salmon showed no significant change
during the spawning run. There were,
however, significant changes in the DDT
concentration of certain tissues.
At present, eggs and fry from Lake
Michigan lake trout are being analyzed for
DDT. After completion of these analyses, a
manuscript covering the results of the lake
trout-coho salmon insecticide study will be
prepared.

UPTAKE OF DDT AND DIELDRIN BY LAKE TROUT,
R.E. REINERT, U.S. Dept. of The Interior,
Bur. of Sport Fish. & Wildlife, Ann Arbor,
Michigan 48107

Seven groups of lake trout were exposed
to different combinations of p,p' DDT and
dieldrin in water and food. The
concentration in water was a few parts per
trillion. The concentration in food was
about 1 ppm. Fish were fed 2 percent of
their body weight per day.
    The accumulation of the insecticides
was followed for 152 days. Fish concentrated
significant amounts of DDT and dieldrin from
food and water. When they were exposed to
DDT and dieldrin in combination, fish
accumulated approximately the same
concentrations of each insecticide as when
they were exposed to them individually.
    After 152 days exposure to the
insecticides was stopped and their breakdown
and elimination was studied. The laboratory
phase of the study has been completed and a
manuscript is being prepared for
publication.

EFFECT OF DIFFERENT POLLUTANTS ON
ECOLOGICALLY IMPORTANT POLYCHAETE WORMS,
D.J. REISH, Calif. State Univ. & Colleges,
School of Letters, Long Beach, California
90801

Primary objective is to study the
effects of selected heavy metals (such as
copper, zinc, mercury & lead), pesticides
(such as DDT & Malathion) and petrochemicals
on the larvae, juveniles & adults of
polychaetes to ascertain the most sensitive
stage in their life history. In addition,
bioassays will be conducted through at least

one life cycle to determine any long-term
effect on reproduction. Ten to fifteen
species of polychaetes will be used in these
bioassay experiments, They will be conducted
on field & lab-reared specimens at all
stages of their life histories. Will also
ascertain the validity of using lab-reared
specimens as test organisms. If results are
similar with both field & lab-reared
specimens, subsequent bioassays will be
greatly facilitated by the use of available
laboratory colonies.

INFLUENCE OF P,P'-DDT AND DIELDRIN ON SOME
MICROSOMAL ENZYMES IN LIVERS OF CHICKENS AND
MALLARD DUCK,
J.L. SELL, North Dakota State University,
Agricultural Experiment Sta., Fargo, North
Dakota 58103

OBJECTIVE: Determine the effects of
feeding p,p'-DDT or dieldrin to chickens and
Mallard ducks on the concentration of
hepatic cytochrome P(450,) the activities of
hepatic hydroxylase and N-demethylase and
the hepatic metabolism of estradiol-17B- C.
APPROACH: Diets containing 0, 10 or 20
ppm dieldrin or 0, 100 or 200 ppm p,p'-DDT
will be fed to White Leghorn pullets and
Mallard duck females of approximately the
same age for eight weeks. Four chickens and
four ducks from each dietary treatment will
be killed after 2, 4 and 8 weeks and liver
samples taken. The activities of aniline
hydroxylase and N-demethylase, the
concentration of cytochrome P(450) and the
hepatic metabolism of estradiol-17B- C will
be determined.
PROGRESS: Two experiments were
conducted to compare the effects of p,p'-DDT

or dieldrin on liver microsomal enzymes of
chickens and Mallard ducks of similar ages.
In experiment one, 41 chickens and 41 ducks
were fed dieldrin at levels of 0, 10 and 20
ppm. The liver samples were obtained at
intervals of 0, 2, 4 and 8 weeks from start
of treatment. It was observed that aniline
hydroxylase and N-demethylase activities of
duck microsomes were 40% and 65%,
respectively, of those of chicken
microsomes, whereas cytochrome P450
concentration in and estradiol metabolism by
duck microsomes were similar to those of
chicken microsomes. Aniline hydroxylase and
N-demethylase activities, cytochrome P450
concentration, and estradiol metabolism were
increased in both species by feeding
dieldrin. In experiment two, 48 chickens
and 48 ducks were fed 0, 100 or 200 p,p'-DDT
and the parameters measured were as
described above. Collection and statistical
analysis of these data are not complete at
this time. An additional experiment was
conducted to determine the response of
Japanese quail to dietary DDT. The
activities of aniline hydroxylase and
N-demethylase were depressed markedly when
DDT was fed at 100 or 200 ppm.
Simultaneously, cytochrome P450
concentration in liver was increased
considerably. Subsequently it was found
that a concentration of 10 M or more of DDT
depressed hydroxylase activity. The
depressing effect of DDT appeared to be by
way of competitive inhibition.

RELATION OF TISSUE PRESERVATION TO RESIDUE
DETERMINATION,
W.H. STICKEL, U.S. Dept. of The Interior,
Bureau of Sport Fish. & Wlfe., Laurel,
Maryland 20810

Birds will be fed a mixture of organochlorine insecticides for 7 days and will then be sacrificed. Hens' eggs will be pooled and mixed with the same pesticides. Identical portions of brains, livers, muscles, and eggs will be preserved by freezing, formalin, or phenoxyethanol. Paired samples will be preserved in different ways to permit direct comparisons. Samples will be analyzed chemically to determine the amounts recovered after different preservations and also to determine the amount of degradation that occurred. Insecticides to be used are endrin, DDT, and heptachlor.

INCREASED REPRODUCTION DUE TO INSECTICIDES, D.J. SUTHERLAND, Rutgers The State University, School of Agriculture, New Brunswick, New Jersey 08903

This project is investigating the possible increase of reproduction, potential and actual, in mosquito and flies due to stress by sublethal amounts of insecticide in the immature stage. Reproduction in the fly has not been increased by DDT and diazinon while it has been in the mosquito, not by diazinon, but by DDT, dieldrin, malathion, Dursban, dichlorvos, and ethanol. This suggests that the mechanism is susceptible to compounds containing chlorine or yielding on metabolism ethanol. Physical stress agents are so far ineffective.

SOME INSECTICIDES IN SOILS AND THEIR DIFFUSION, INTO PLANTS, SOIL MACROFAUNA AND GROUNDWATER, A.W. TAYLOR, Inst. of Plant Protection, Poznan, Poland

OBJECTIVE: Measure the accumulation in soil, uptake by plants and decomposition rates and products of selected insecticides, and determine the effects of these insecticides on the populations of earthworms, wireworms, grubs and other soil fauna.

APPROACH: The insecticides DDT, Carbaryl and Fenitrothion tagged with C-14 and other isotopes will be applied to 3 m soil areas on which alfalfa, small grains, or vegetables will be grown. Residues of the insecticides and their decomposition products in soil and plants will be identified and determined. The effect of the insecticides on populations of and accumulation by soil fauna will be evaluated.

SELECTION OF DROSOPHILA FOR RESISTANCE TO INSECTICIDES,
C.H. THOMAS, Mississippi St. University, Agricultural Experiment Sta., State College, Mississippi 39762

OBJECTIVE: Determine the amount of insecticides required to reduce the number of progeny in strains of Drosophila and the type of inheritance involved. Produce strains that are resistant to one or more insecticides. Determine if a resistant strain will become susceptible if selection ceases.

APPROACH: Drosophila will be raised separately and in combination on medium containing sufficient quantities of insecticides to reduce the number of progeny. After the quantity of insecticide required to reduce the number of progeny has

233

been determined, that amount will be used each generation. Progeny will be selected at random from each insecticide and control group each generation and reproduced on medium with and without insecticides. Crosses will be made between the lines selected for resistance to different insecticides and their progeny will be tested for degree of resistance.

PROGRESS: For 25 generations 5 strains of Drosophila have been reared on media containing 4 levels of the insecticides Sevin, DDT, Dieldrin and Malathion. The smallest and largest concentrations of insecticides expressed in ppm between generations 11 and 25 to reduce the progeny 0% in the lowest level and 90% in the highest level as compared to a control and the lowest and highest level in generation 25 were 1.8 in h and se strains and 69 in B1L/Cy strain and 1.8 in h and se and 69 in B1L/Cy for DDT; .01 in vg and 1.1 in B1L/Cy and h and .05 in B1L/Cy and .5 in B1L/Cy, h and se for Dieldrin; .02 in B1L/Cy and 4 in wild and .05 in B1L/Cy and 1.2 in wild, h and se for Malathion; .8 in se and 71.4 in wild and se and 1.6 in se and 71.4 in wild and se for Sevin. Reciprocal crosses were made between flies of the same strain obtained from the highest concentration of the 17th generation of Dieldrin and the 16th generation of Malathion and between flies from the 18th generation of Sevin and the 19th generation of Dieldrin. There was a highly significant difference between strains and media in the F(1) and F(2) and between reciprocal crosses in the F(2) of the Dieldrin X Malathion crosses and between the strains and media in the Dieldrin X Sevin crosses. When progeny from files that were produced on insecticides from generations 11 to 25 were allowed to reproduce on media without insecticides, they produced as many or more progeny than the controls in most tests.

BIODETERIORATION OF NAVY INSECTICIDES IN THE
OCEAN,
H.P. VIND, U.S. Navy, Civil Engineering Lab.
, Port Hueneme, California 93041

DETERMINE THE RELATIVE RATES AT WHICH
VARIOUS INSECTICIDES USED BY NAVY PEST
CONTROL OPERATORS ARE DETOXIFIED IN
SEAWATER. ISOLATE MICROORGANISMS WHICH WILL
DEGRADE DDT. OBTAIN A SUBSTITUTE FOR DDT
WHICH IS MORE EASILY BIODEGRADED.
BIODEGRADABILITY OF INSECTICIDES IN THE
OCEAN WILL BE EVALUATED BY ASCERTAINING THE
RATES AT WHICH INSECTICIDE-IMPREGNATED
MATCHSTICKS LOSE THEIR ABILITY TO WARD OFF
MARINE WOOD-BORING CRUSTACEANS. TESTS WILL
BE RUN TO ELIMINATE LEACHING AS A CAUSE FOR
DECLINE IN TOXICITY. IN OTHER EXPERIMENTS,
ATTEMPTS WILL BE MADE TO CULTIVATE BOTH
ANAEROBIC AND ANEROBIC MARINE BACTERIA ON
VARIOUS TYPES OF CHLORINATED HYDROCARBONS.
RESIDUAL CARBON-BOUND CHLORINE WILL BE
DETERMINED FOR EACH TYPE AT THE END OF TWO,
SIX AND TWELVE MONTHS.
SUPPORTING AGENCY ADDRESS INFORMATION:
OFFICE OF NAVAL RESEARCH 443 ARLINGTON, VA.
22217

EFFECT OF ENVIRONMENTAL POLLUTION ON FISH
DISEASES - ORGANOCHLORINE PESTICIDES AND
DISEASE RESISTANCE MECHANISMS OF RAINBOW
TROUT,
G. WEDEMEYER, U.S. Dept. of The Interior,
Bur. of Sport Fish. & Wildlife, Seattle,
Washington 98115

Inhibition of Na ion, K ion-activated
adenosinetriphosphatase (salt pump) of
kidney, brain and gills is potentially
important in the susceptibility to stress,
and hence infection, which has been noted in

fishes chronically exposed to organochlorine
pesticides.

The hypothesis will be investigated
using chlordane, DDT, and endosulfan with
the rainbow trout as the test species.

INTERCEPTION AND DEGRADATION OF PESTICIDES
BY AQUATIC ALGAE,
L.R. WORTHEN, Univ. of Rhode Island, School
of Pharmacy, Kingston, Rhode Island 02881

The proposed research involves both
collecting blue-green algae from several
stations over a two year period and the
propagation of axenic cultures of several
representative species.

Those collected from natural sources
will be examined to determine the extent of
pesticide (DDT and dieldrin) accumulation.
Radioactive isotopes of the two pesticides
will be used to determine if absorption of
these compounds takes place, and if these
compounds are accumulated in the organism,
or more important, if they are degraded and
the pesticide metabolites formed.

THE EFFECT OF DDT ON EGG, FRY AND FINGERLING
VIABILITY IN F2 STRAINS OF BROOK TROUT,
J.E. WRIGHT, State Fish Commission,
Harrisburg, Pennsylvania 17120

Objective: To select those brook trout
strains which are most resistant to harmful
effects of DDT and its metabolites
accumulated in the tissues and eggs.

Procedure: Females of control and
treatment lots described in Job 1, Study 1,
will be spawned individually and mated to

individual sibling males from control and
treated lots. Samples of unfertilized eggs,
as well as each spawned fish, will be
analyzed for insecticide levels as described
in Job 1 Study 1. Total egg number, size,
fertility and viability records will be
obtained for the individually spawned lots
through the fry stage. Those progeny lots
will be saved for breeding stock which show
significantly higher viability.

REVIEW OF EXISTING LITERATURE AND
PUBLICATION OF FINAL REPORT ON DDT IN BROOK,
BROWN, RAINBOW AND LAKE TROUT AND COHO
SALMON,
J.E. WRIGHT, State Fish Commission,
Harrisburg, Pennsylvania 17120

Objective: Review and summarize
available material on the accumulation of
DDT and its metabolites in brook, brown,
rainbow and Lake trout as well as coho
salmon.
Procedure: Review files in Benner
Spring Research Laboratory, Pennsylvania
State University Library and other libraries
for applicable material and prepare summary
reports of material that can be utilized.

THE EFFECTS OF DDT ON COLLEMBOLA FAUNA OF
MEADOW SOILS IN YUGOSLAVIA,
J. ZIVADINOVIC, Univ. of Sarajevo, Sarajevo,
Yugoslavia

OBJECTIVE: Investigate the effects of
DDT on soil fauna, especially Collembola.
APPROACH: Experiments designed as
block-square with 3 treatments and 1

237

untreated control, each replicated 3 times,
will be conducted on two meadows, one
cultivated and one uncultivated. Each
treatment square will receive a different
concentration of DDT applied twice annually.
Soil samples will be taken from all squares
once a month for 2 years with a standard
5-x10-cm. cylinder. Apterygota and
Myriapoda will be separated by Berlese
funnels, preserved, and identified. The
effects of DDT on the soil fauna will be
assessed, as will measurable changes in soil
quality.

71001